# THE NATURAL WORLD

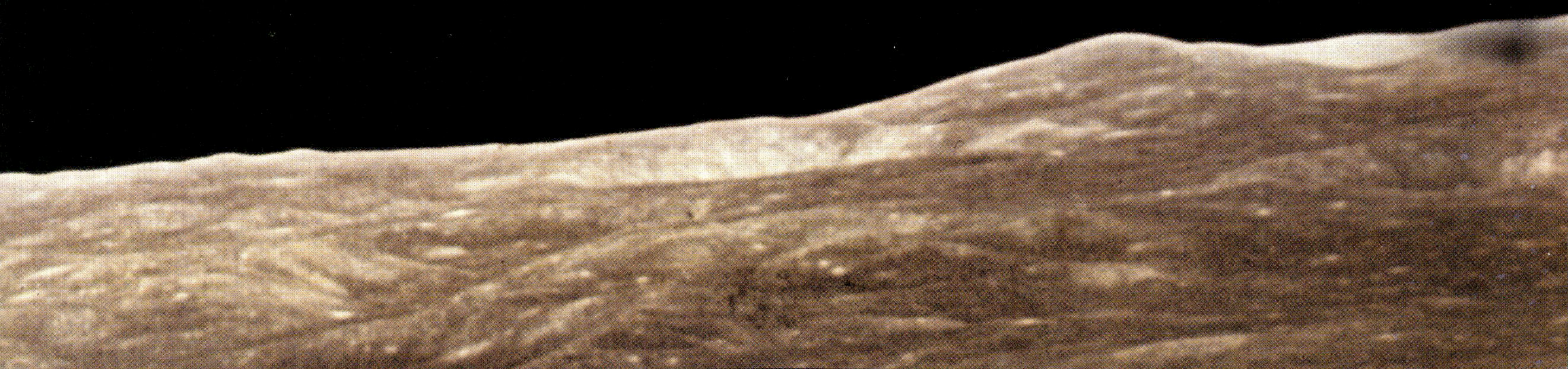

HOW THINGS WORK
SYSTEM EARTH
LIBRARY OF CURIOUS AND UNUSUAL FACTS
BUILDING BLOCKS
A CHILD'S FIRST LIBRARY OF LEARNING
VOYAGE THROUGH THE UNIVERSE
THE THIRD REICH
MYSTERIES OF THE UNKNOWN
TIME-LIFE HISTORY OF THE WORLD
FITNESS, HEALTH & NUTRITION
HEALTHY HOME COOKING
UNDERSTANDING COMPUTERS
THE ENCHANTED WORLD
LIBRARY OF NATIONS
PLANET EARTH
THE GOOD COOK
THE WORLD'S WILD PLACES

This volume is one in the Time-Life series
HISTORY OF THE WORLD.

# THE NATURAL WORLD

**BY THE EDITORS OF TIME-LIFE BOOKS**

TIME-LIFE BOOKS, AMSTERDAM

**TIME-LIFE BOOKS**

EUROPEAN EDITOR: Ellen Phillips
*Design Director:* Ed Skyner
*Director of Editorial Resources:* Samantha Hill
*Chief Sub-Editor:* Ilse Gray

**TIME-LIFE HISTORY OF THE WORLD**

Editorial Staff for *The Natural World*
*Editor:* Charles Boyle
*Designer:* Lynne Brown
*Researchers:* Lesley Coleman, Celia Dearing, Deborah Pownall
*Sub-Editors:* Frances Willard, Luci Collings, Tim Cooke
*Design Assistant:* Sandra Archer
*Editorial Assistant:* Molly Sutherland

Picture Department
*Picture Administrator:* Amanda Hindley
*Picture Coordinator:* Elizabeth Turner

Editorial Production
*Production Assistant:* Emma Veys
*Editorial Department:* Theresa John, Juliet Lloyd-Price


First European English language printing.


ISBN 0 7054 0992 9


**CONSULTANTS**

**General:**
NEIL C. ROBERTS, Lecturer in Physical Geography, Loughborough University of Technology.

**Classical:**
GRAEME BARKER, Professor of Archaeology, University of Leicester.

**China:**
CHRISTOPHER CULLEN, Lecturer in the History of Asian Technology, School of Oriental and African Studies, University of London.

**Medieval and Early Modern:**
DENIS COSGROVE, Reader in Cultural Geography, Loughborough University of Technology.

SPECIAL CONTRIBUTORS: Neil Fairbairn, Ellen Galford, Michael Kerrigan, Alan Lothian, Clio Whittaker (text).

Correspondents: Elisabeth Kraemer-Singh (Bonn); Maria-Vincenza Aloisi (Paris); Ann Natanson (Rome). Valuable assistance was also provided by: Louise Forstall (Alexandria, Virginia); Yang Jinquan (Beijing); Elizabeth Brown (New York); Josephine du Brusle (Paris); Traudl Lessing (Vienna).

# CONTENTS

# OUT OF THE WILDERNESS

1

Trekking across the desert wastes of the Rub al-Khali in Saudi Arabia, a party of Bedouin nomads accompanied by an English explorer advanced in silence beneath the sun's relentless glare. They progressed at a steady walking pace to conserve the energy of their camels: the only food for these beasts were the dry leaves and thorns of occasional bushes, and it might be several days before the party reached the next water hole. On the flat gravel plains they averaged around five kilometres an hour; where the dunes were steep and the sand yielded at every footstep, they managed no more than half this rate. They had seen no other human beings for 17 days.

Abruptly the party came to a halt. A grey-bearded Bedouin dismounted and examined some partly blurred camel tracks on hard ground. "Who were they?" demanded the group's leader. The bearded man broke some camel droppings in his fingers and said that the tracks had been made 10 days before by six members of a particular tribe; the riders had raided another tribe on the southern coast and had stolen three of their camels; they had last taken water at a well named Mughshin. The Englishman, Wilfred Thesiger, had learnt not to be surprised. "Here every man knew the individual tracks of his own camels," he reported, "and some of them could remember the tracks of nearly every camel they had seen. They could tell at a glance from the depth of the footprints whether a camel was ridden or free, and whether it was in calf. By studying strange tracks they could tell the area from which a camel came."

For the Bedouins, reading imprints in the sand with such accuracy was as commonplace a skill as is, say, a contemporary city dweller's ability to drive a car or to book a theatre ticket by telephone. It was also crucial to their survival: it could tell them the whereabouts and proximity of their enemies, a precious water hole or an animal for the cooking pot. And this was but one example of their intimate cognizance of the harsh environment in which they lived. Their hereditary knowledge had enabled them to feed and clothe themselves and cure their ailments by methods that had hardly varied for thousands of years.

Wilfred Thesiger travelled with the Bedouins for many months in the late 1940s, and the qualities he came to admire most in his companions were their dignity, courage, endurance and generosity. But when in the 1970s the oil wealth of Saudi Arabia transformed for ever the lives of the Bedouins, when trucks replaced camels and television aerials sprouted from black felt tents, what the rest of the world had most to mourn was the loss of an entire memory bank of knowledge of the natural world. When skills become no longer needed they quickly die out.

Three thousand years ago the knowledge possessed by the Bedouins was akin to that of all other human communities inhabiting the earth. Most of these people lived by one of three overall strategies: they survived by hunting animals and gathering plants, by simple methods of agriculture, or — like the Bedouins — off the herds of

**Painted probably some 4,000 years ago on a rock face in the region of Tassili N' Ajjer — "plateau of the rivers" — in southern Algeria, a hunter launches his spear towards a horned wild sheep. Snapping at the heels of the prey is a dog, the first animal species tamed by humans. Between around 7000 and 3000 BC much of the now arid Sahara was well-watered grassland inhabited by humans who gathered edible plants and hunted herds of wild game, including antelopes, giraffes, ostriches and elephants. Some regions also yielded freshwater turtles, molluscs and fish.**

Among the sharp, dry grass and scraggly trees of the Kalahari Desert in southern Africa, crouching hunters of the !Kung people keep downwind of their prey. Their bows and arrows hardly differ from those used by their ancestors from around 10,000 BC. The spears which they carry to finish off a wounded animal have an even longer pedigree: flaked points such as the 40,000-year-old example on the right – shaped to slot into a shaft of wood or bone – were fashioned from around 125,000 BC, enabling hunters to kill with new and deadly efficiency.

domesticated animals migrating across huge tracts of land in search of fresh pastures. Each of these life styles originally evolved in response to particular environmental conditions, and for countless generations they proved capable of satisfying all basic human needs for food and shelter.

Later and more populous societies were to become more exacting in their needs, more specialized in their use of natural resources and more vulnerable in times of crisis. Successive civilizations inhabiting the Mediterranean region – culminating in the awesome grandeur and power of the Roman empire – discovered that the progressively more efficient exploitation of nature could reap diminishing returns. In China and other parts of Asia, the needs of fast-growing human populations tested the limits of human skill and ingenuity. The Western world, moving by degrees from levels of bare subsistence to unparalleled affluence, solved its own problems of food supply by the application of advanced scientific and technological processes – and came late to a realization that short-term gains can be offset by long-term losses on a far greater scale. But the experience of humans who lived before cities were developed has never lost its relevance, for they too were not immune to ecological crises, and the pattern of their development followed a now familiar course.

Even as they roamed vast areas hunting mammals and birds, catching fish, and gathering nuts, berries and other plant foods, the earliest humans were beginning to shape the development of the world they lived in. Having come to know the ways of this world so intimately that they could anticipate where and when a particular plant could be expected to yield its richest harvest, and where migrating herds might most easily be ambushed, they proceeded to intervene directly in its workings – by burning wooded land, for instance, to encourage the growth of favoured food plants. It was only a matter of time before humans began to cultivate plants as the foundation of their economic existence. Settled in one place and growing their own food, they were able to accumulate a surplus with which to attract and tame wild grazing animals – sources of meat, wool, hide and haulage. In some drier regions, communities that proved successful in stock-raising eventually took to the nomadic life once more – this time as pastoralists, driving their herds from one area to another, often over great distances, to take full advantage of the seasonal growth of pastures.

Such specialized life styles provided sustenance, and at times even affluence. Yet these achievements were bought at a high price: the more control humans gained over their environment, the more serious and lasting was the impact they had upon it; at the same time, however, the more they came to depend upon single sets of resources, abandoning the opportunistic flexibility of earlier hunter-gatherers, the more immediate and severe was the impact of any environmental change upon them.

Fortunately, because potentially of great benefit to human understanding of the natural world, there still exist in certain remote and inhospitable parts of the globe peoples who maintain with their habitat a traditional relationship that is relatively undisturbed by the political and economic imperatives of modern civilization. These groups comprise a minute proportion of the global population; but their life styles are palpable evidence of how all humanity once lived, and of a closeness to nature that the majority of humankind has lost.

The world into which humans first emerged was itself in the grip of profound and continuous change. The shape of the land was being slowly moulded by the effects of a succession of great freezes, each one lasting for about 100,000 years, which were interrupted by shorter and milder periods. Massive ice sheets — covering what is now Canada and Scandinavia — crept gradually southwards century by century; at times as much as five kilometres thick, they carved out new landscapes as they scraped and gouged their way across the northern continents. Both northern and southern ice sheets drew moisture from the earth's atmosphere; as a result, sea levels fluctuated sharply, dropping by more than 100 metres during the most frigid periods and re-drawing the map of the world. Asia and North America, separated in modern times by the Bering Strait, were joined by a wide, low-lying land bridge across which both animals and humans were able to make their way. Far over to the west, meanwhile, the British Isles, together with a large area of the North Sea, formed a continuous promontory of continental Europe.

While the homelands of the first human beings in southern Africa remained sub-tropical in climate throughout even the most frigid periods, the Eurasian areas in which modern humans had begun to settle some 40,000 years ago were bearing the full brunt of the great glaciations. There, where the bleak, sterile expanse of the ice sheet ended the tundra began — a vast and monotonous terrain that no summer ever thawed beyond the top few centimetres of soil. Farther south, a broad expanse of grassland or steppe stretched across the breadth of Eurasia. Burgeoning in a riotous blaze of colour during the few short weeks that constituted summer each year, its millions of square kilometres of grass — sparse and tussocky as it was — was home to large herds of grazing mammals, including reindeer, horses, mammoths, woolly rhinoceroses and musk oxen. Beyond the steppe lay forest refuges, and from here, as the ice sheets slowly receded during the periodic milder periods, the leafy, deciduous woodland pushed northwards.

In western Europe, which was blanketed by the deep waters of the Atlantic Ocean, the winters were considerably milder and the summers cooler than those experienced at the heart of the Eurasian landmass. Between 30,000 and 12,000 years ago this area was home to significant numbers of humans who took their supplies of meat from the rich game reserves the continent offered, and gathered a variety of vegetable food from a wide range of plants. Living mostly in small bands numbering around 25 people, they would regularly join up together into larger tribal groups in order to mount communal hunts with greater speed and efficiency, or to sit out the cold, dark months of winter in caves. Throughout the spring and summer months, on the other hand, when the verdant, blooming steppe provided an abundance of easily gathered plant foods, these bands might once more fragment into smaller family units that comprised simply a man, a woman and their children.

The size of the social group was very closely determined by the available food supply. Members of the group cooperated closely, sharing both food and leadership, though by and large operating a very clear sexual division of labour. Men hunted animal foods, and the women gathered the vegetable products — nuts, berries, and edible leaves and grasses — as well as birds' eggs, small mammals and, by rivers and coasts, shellfish.

Roaming at any given time in a relatively confined block of land of a few hundred square kilometres, within a far larger territory of up to 4,000 square

**A rock painting in Spain shows angry bees swarming round a precariously balanced woman as she reaches into their nest to extract honey. Among the foodstuffs gathered by women — which included plants, roots, berries, insects and birds' eggs — honey was a prized delicacy, and was eaten together with the larvae and the comb. Some contemporary hunter-gatherers make a mildly alcoholic wine by squeezing the honey from the comb, dissolving it in warm water and adding yeast, then leaving it to ferment for a day and a night.**

kilometres, a band of hunter-gatherers would periodically venture farther afield to occupy a new block. Each area of land was thus given a chance to recover after a few years' exploitation, while the group remained familiar with the full range of resources offered by its territory as a whole, without ever becoming overdependent upon the products of one particular zone. Dispersed in this manner over hundreds of square kilometres, the hunter-gatherer band was able to keep its impact on its environment to a minimum. The impact was, for all that, by no means negligible.

During colder periods the dietary mainstay of the early European hunter-gatherers was provided by the herds of large mammals such as bison, red deer, horses, reindeer and mammoths that grazed the open steppe. Milder periods saw an increase in the importance of smaller, woodland species such as roe deer and wild pigs. Hunter-gatherers living in warmer climates tended to rely more heavily upon vegetable foods, which in the tropics were abundant throughout the year, and easy to find, if rather laborious to gather. But the growing season in western Europe at cold times was very brief, and the women in a group had to be ready and able to respond extremely quickly and with careful planning if they were to exploit its harvest to the full. The important plant varieties would not grow alongside each other, but would be scattered in different habitats across a wide area, and for each to be caught exactly at its moment of ripening required forethought and organization.

Moreover, the considerable nutritional value that was offered by plant foods and such gathered animal foods as birds' eggs and shellfish was to some degree offset by the labour required in their gathering and preparation. While it might take a group of hunters a great deal of time, energy and ingenuity – not to mention luck – to hunt down a red deer, for example, that animal once killed would provide the calorific equivalent of more than 52,000 oysters, each of which would need laborious picking and shelling before its nutritional possibilities could be realized. Given the abundance of grazing animals to which the hunter-gatherers had ready access, it was inevitable that they would tend to rely quite heavily upon meat in their diet. Animals provided a range of other products too: from hide, fur and sinew for making blankets and sewn clothing, to bone and ivory for tools, weapons and – where wood was scarce – for use in bulk as a building material and as a fuel.

From around 20,000 years ago, humans developed technologies that made hunting significantly easier, as well as less dangerous than it had been previously. For earlier generations of hunters, armed with only their weighty and cumbersome stone clubs, spears and crude axes, the hunt had been a rough and gruelling battle between man and beast, fought out at close quarters at considerable risk to the man. Now, however, the bow and arrow were introduced, and the spear-thrower was added to the hunter's kit. This latter device was a length of wood or bone about half a metre long, with a hook at one end which clipped over the butt end of a spear. The spear-thrower acted as an extension of the human arm: held back over the shoulder, then snapped sharply forwards, it sent the spear flying swiftly through the air towards its target, straight and true. No longer just a crude stabbing weapon, the spear now became a deadly missile that could be accurate up to a distance of around 30 metres. Fine stone blades, mass-produced by expert craftsmen from prepared flint cores, and delicate slivers of bone cunningly fashioned into lethal points added to the impact of both arrows and spears. Hunting was becoming less a dangerous, physical brawl between hunters and hunted and more an art that depended to a large extent upon skill and finesse: a battle of wits in which man's advantages were clear.

Along with the new weapons came changes in tactics. Cleverly planned ambushes and game drives now allowed bands of hunters to kill in huge numbers, making them far more dangerous than any animal predator had ever been. Siting their encampments beside river fords, the hunter-gatherer bands positioned themselves to attack herds of migratory mammals at a vulnerable moment during their annual journey. Migrating mammoths might be ambushed where a gently sloping hillside offered a convenient butchering floor upon which the vast carcasses — far too heavy to be dragged any distance — could be carved up where they fell, and on which the meat obtained could then be dried in the sun.

Fire, which was first used by humans to scare off predatory animals and to cook food, was also employed in hunting. Not all fires were started by humans; many must

## THE DIVINE RADIANCE

**Until the world was manipulated to yield other sources of energy, all human beings were utterly in thrall to the sun, the foundation of every form of life that they encountered. The growth, survival and reproduction of plants is dependent on the sun's radiant energy, which is trapped by their green leaves; when plants are consumed, the same energy powers movement and promotes the growth of body tissue. The images shown on these pages and overleaf illustrate the high position accorded to the sun in the pantheons of early societies — an intuitive acknowledgement of the great debt that they owed to the golden disc that slowly swung across their skies, and whose nightly vanishing intimated their mortality.**

**In a stone stele of the ninth century BC, Shamash, the Mesopotamian sun-god, receives homage in his shrine; his emblem, a solar disc, rests on an altar.**

**Zigzag beams radiate from a beaten gold mask of Inti *(opposite page)*, the supreme god of the Inca people of Peru from around AD 1200.**

have occurred naturally, caused by lightning. But the deliberate use of fire in hunting was sufficiently widespread to have a major impact on the environment. In some regions, when a patch of woodland or scrub was burnt, man discovered that the extra sunlight and the nutrient-rich layer of ash that resulted at ground level encouraged the growth of food plants – whether for humans to eat themselves or as a means of attracting game animals. On the other hand, persistent burning of trees and undergrowth left the topsoil unprotected and liable to be easily washed away by any subsequent rain showers, especially on sloping ground.

Setting undergrowth ablaze both improved sightlines and drove any sheltering animals out into the open, where waiting hunters could be ready to attack with spears and bows and arrows, or with nets for smaller animals. More deadly still, fires were set to startle grazing herds and direct their stampede towards cliffs or steep banks. Plummeting in their hundreds to the bottom of deep ravines, the terrified animals were subsequently finished off by a hail of spears and arrows, or by rocks showered down on them from above. Such strategies yielded meat for the hunters in hitherto undreamt of quantities: at the foot of one particular cliff in France, for example, the remains of more than 100,000 wild horses have been found. Much of the meat obtained in this way was wasted, since the hunters often ended up with more than

they could use or conveniently store, and animals lying buried underneath piles of heavy carcasses were often simply left to rot.

Hunting on a near-industrial scale, using ambushes and fire, probably contributed between 14,000 and 10,000 years ago to a wave of extinctions, notably among large herbivores such as the mammoth, woolly rhinoceros, giant deer and American horse, and also among the great carnivores such as the sabre-toothed tiger that had preyed upon them. The effect was most pronounced in the Americas where, within just a thousand years of the arrival of humans across the land bridge from Asia, dozens of varieties of large mammals died out. Changes in climate and vegetation also played their part – in particular, the advance of woodland into the grassland habitats favoured by the larger herbivores. On the other hand, when such drastic transformations had last swept the earth – during a great warming that had commenced some 120,000 years earlier, and before the arrival of human hunters on the scene – there had been no comparable cost in animal species. The picture is further confused by the fact that those animals known to have been most commonly killed in large-scale drives, such as the reindeer in Eurasia and the bison in North America, survived in abundance into modern times; many of Africa's large species also came through largely unscathed, despite hunting by man. The real reasons for the extinctions remain in doubt, although it is clear that humans had become powerful enough to at least influence developments in the natural world.

For all the efficiency of the new weapons and tactics, however, the relationship between the hunters and the hunted encompassed the spiritual as well as the pragmatic. Most hunting expeditions were still close-run affairs in which the wits and cunning

**The round emblem painted on this limestone tablet in Sweden in the fifth century AD represents the sun, whose life-giving powers were invoked by early Norse peoples at midwinter festivals.**

**A wooden shield made probably in the 18th century bears the visage of the Indian sun-god Surya, worshipped from 1500 BC until the present day.**

of humans and animals were evenly matched. To have any chance of success, the hunter had to possess a very thorough understanding of his prey's natural behaviour; this understanding extended to respect and empathy, and partook of the hunters' religious sensibilities concerning the other forms of life with which they shared the earth. These feelings were given tangible form in an immense outpouring of artistic activity: many thousands of personal ornaments and other artefacts have been found, which must represent a mere fraction of the output of the period. Most are small and portable, in keeping with the nomadic hunter-gatherer existence: fine necklaces and beads fashioned from colourful stones and seashells, and remarkable human and animal figures worked in stone and clay, bone and ivory. More permanent works survive in the form of paintings executed on the walls of caves in which the hunters sheltered. The most celebrated surviving examples are at Lascaux in France, and Altamira in northern Spain: here, vividly captured in paints prepared from charcoal and coloured earths, bison, wild bulls, stags and horses buck and charge across the cavern walls, the flat images given depth by the contours of the rock, from which they appear to grow organically. Around these leaping forms, mysterious symbols complete larger designs. The meaning and purpose of these paintings – whether they played a part in fertility rites, represented favoured totemic animals revered as protectors, or perhaps were simply decorative – cannot be known, but the feelings of awe that inspired their execution are still apparent.

Far to the south, similar paintings were produced by hunter-gatherers in caves along South Africa's Drakensberg, a craggy escarpment snaking through Natal, Lesotho and into the remote eastern part of Cape Province. And here, in territory generally considered useless by later civilizations, the descendants of the original artists have survived. These people are the !Kung – the exclamation mark represents a click of the tongue, much used in their language. (They were formerly known as the Bushmen, an appellation which in South Africa acquired derogatory associations.) Their modern domain lies some distance to the north of the Drakensberg, in the western part of the Kalahari Desert, a parched, sunbaked expanse of scrub and waving grass punctuated by occasional trees, with little in the way of surface water, covering thousands of square kilometres of southern Africa's interior. No respecters of modern political boundaries, the !Kung bands roam a large area that straddles the border between Botswana and Namibia.

Though the Kalahari is by no means completely arid, its rainfall is erratic, and much of a band's energy is devoted to finding water. Long-standing and ritualized habits govern the use of this precious resource: the more obvious and lasting sources are left until seasonal, less easily exploited supplies are exhausted. Hunters who are travelling far from their camp, or from low-lying land that is likely to have water available on the surface, carry water with them in the empty shells of ostrich eggs. When this supply runs out a hunter can use a straw to reach water trapped in hollow trees or root systems, or suck water from beneath the sand through a tubular stem.

The writer Laurens van der Post, who was born on the edge of the Kalahari in 1906, once witnessed a hunter named Bauxhau extract water in this way. The hunter first dug a hole in the sand about an arm's length deep. He then took the stem of a bush with a hollow core and wound some grass around one end, to serve as a filter. He inserted the stem into the hole and packed sand tightly around it, then he arranged some empty ostrich eggshells by its side. Finally he took a short stick and placed one end in his mouth and positioned the other over one of the shells. When all of these preparations were complete, he put his lips to the tube. "For about two minutes he sucked mightily without any result," reported van der Post. "His broad shoulders heaved with the immense effort and sweat began to run like water down his back. But at last the miracle happened and so suddenly that I had an impulse loudly to cheer. A bubble of pure bright water came out of the corner of Bauxhau's mouth, clung to the little stick, and ran straight down its side into the shell without spilling a drop!" When all the shells had been filled, van der Post himself attempted to repeat what the hunter had done, but could extract no moisture at all.

The !Kung's intimate knowledge of their environment is as essential to their hunting and gathering of food as it is to their finding of water. They take their sustenance from around 100 different animal species and 150 plants. Of particular importance is the mongongo nut, the fruit of trees that grow in groves in the sandy soil around rock outcrops. These highly nutritious nuts, which are protected from rotting after they have fallen to the ground by their hard shells, yield almost half of the calories consumed by the !Kung in their vegetable diet. The food that the !Kung prize most is the meat of large antelopes such as the kudu, gemsbok and eland, but usually they make do with smaller antelopes and wart hogs, and also birds — trapped or brought down with arrows — and their eggs. Some species of lizard and snake are also eaten. Insects — apart from a few special delicacies — are not much valued, but the honey of wild bees is a very popular treat. One particular type of beetle performs an important function: it yields a venomous fluid that is fatal when absorbed into the bloodstream but harmless when ingested orally. Applied with care to arrowheads, it turns small darts into lethal weapons, while leaving the resulting meat quite safe for human consumption.

In general, plant foods, abundant all year round in the warm climate of the Kalahari, form the staple of the !Kung diet, and animal protein is less important than it was for the early European hunter-gatherers. Though the !Kung do what they can to nurture grazing animals, burning large areas each winter to encourage the growth of new grass, game in the modern Kalahari is not as plentiful as it once was in Europe, and hunting is more difficult and time-consuming. Each animal kill, it has been estimated, requires an average of four man-days' hunting. Even after being hit by a poison arrow, a large antelope may still have to be tracked for several days before it can be finished off. Perhaps in response to the relative scarcity of big game, the !Kung have evolved a taboo upon killing more meat than is necessary. They also tend to pick upon surplus males among the animal population.

In most regions of the world the hunter-gatherer way of life gradually gave way to other forms of activity more appropriate to settled and larger communities. But it

was the norm for well over 90 per cent of humanity's time on earth, and the system gave those who practised it – as it still gives the !Kung – a simple yet adequate standard of living in return for a relatively small investment of effort. In the case of the !Kung, it has been estimated that a leisurely 2.4 working days per week are enough to provide for their needs. In addition, the nomadic hunter-gatherers' extensive knowledge of the resources available in their territory, their physical mobility and their dietary flexibility in responding to local gluts and shortages, have enabled them to ride out unpredictable natural disasters and climatic fluctuations rather more easily than could their settled, specialized successors.

By 10,000 years ago there were some 10 million people in the world. Small as this figure seems by later standards, it represented, given the extensive territories required for the hunter-gatherer existence, the beginning of a crisis in population density. More intensive ways of reaping the earth's resources were required. The deliberate cultivation of crops and husbanding of animals would meet this need – yet it would lack the light touch of the hunter-gatherer ways. In making the earth work that much harder, agriculture was to have a radical and enduring impact upon it.

Hunter-gatherers themselves gradually began to control the growth of favoured food plants, taking seeds from natural patches of vegetation and planting them in more convenient and more productive sites. Over the centuries human intervention increased, with dramatic consequences. Plants taken into cultivation, harvested and then resown year after year, tend to change genetically in response. The main seed-bearing core of a wild plant is usually brittle, encouraging propagation by allowing the ripened seeds to scatter at the slightest breeze. The specimens favoured by humans were those whose cores were tougher than average, and therefore more likely to retain their seeds long enough to be reaped by the farmer's sickle. Through several generations the resilient core became a standard, genetically ordained feature of the plant in its cultivated form – so much so that the plant became wholly domesticated, unable to release its own seeds, and dependent instead upon its cultivators' stripping the grain from the stalk and planting it in the ground. Similarly, when farmers came to select the seed for each year's sowing, they chose the largest, plumpest grains; over time, as a result, average size increased and the domesticated plant grew into a distinct form, often bearing little apparent resemblance to its wild ancestor.

Once plants had been taken under human control and were producing far higher yields, some of the surplus generated could be used for feeding animals, thereby accelerating their own domestication. Again, through many generations of human

**The raging forest fire shown in this artist's drawing represents the most powerful impact of hunter-gatherers upon their landscape. As well as using fire to cook food and harden the tips of their weapons, hunters learnt to set undergrowth alight to drive animals into the open or stampede herds towards a prepared ambush, and also to encourage the regrowth of tender young plants that attracted animals to graze. Ground temperatures in forest fires can reach 150 degrees Celsius, and over time the recurrent effects of such heat caused the local extinction of certain plants while favouring those with thick protective barks or those that could regenerate from underground parts.**

The flare of ripening corn against the waters of the Euphrates River in present-day Iraq proclaims the human ascendancy over nature empowered by agriculture. Domesticated cereal crops, first flourishing in the Fertile Crescent in the eighth millennium BC, were soon seeded along the banks of the Nile in Egypt, by the River Indus in Pakistan and in the valleys of Central America. The labour of the humans that they supported was changed from pursuit and search to planting and harvesting, as shown on the right in an Egyptian tomb painting dating from 1300 BC. Overleaf, an Egyptian limestone statuette represents a maidservant grinding corn into flour: bowed to new rhythms, the human body learnt new aches.

husbandry, domesticated animals grew into distinct forms, so genetically altered that they could no longer reproduce with their wild relations. Unlike plants, animals usually became smaller in their domesticated forms. Size differences between males and females were eroded, and adult animals tended to retain juvenile characteristics of bone structure and colouring – piebald patterning, for instance, common in domestic animals and young wild ones, is unusual in adult wild animals. Domesticated animals provided not only meat, hide, wool and hair, but also dung for fertilizing the soil and labour for drawing ploughs.

In a range of environments across the world, a balanced selection of both plants and animals was brought under human management. In the Fertile Crescent in the Near East, early forms of cereals such as wheat, barley and rye, and pulses such as lentils and peas, were farmed alongside animals including cattle, sheep and goats. Irrigation allowed the gradual outward spread of agriculture into the drier steppe areas of the Eurasian interior. In China and Southeast Asia rice and millet provided the staple grains, and were cultivated with the assistance of strong, domesticated water buffaloes. Across the Pacific, farmers in South and Central America grew root crops: potatoes, yams and cassava. In the high Andes llamas and alpacas, carrying their heavy loads up and down the vertiginous mountain tracks, helped to make large-scale cultivation practicable.

Farther north, through much of Central America from the lush rainforests of the coastal plains to the sunbaked hill country of the interior, a different system of agriculture was slowly taking form. Some 7,000 years ago, in the hot, dry Oaxaca valley of southern Mexico, peoples who had lived for countless generations by hunting white-tailed deer and cottontail rabbits on grassy hillsides, catching mud turtles in ponds and creeks, and gathering a wide range of vegetable foods, began to cultivate some of their most important plants as a stand-by to supplement what they could manage to gather from the wild. Their squashes, beans and teosinte – a seed-bearing local grass – began to evolve into distinct strains, steadily increasing in size. Its seed-heads growing from less than three centimetres in length to over 20, teosinte became the domesticated cereal known in modern times as maize.

For many hundreds of years crop growing in the Oaxaca valley remained a supplementary activity, and it was not until some 3,800 years ago that the evolving

maize cob had grown to a size that allowed the Oaxacans to make agriculture their main pursuit. This relatively slow rate of progress was to a great extent due to the absence in the region of large animals that were suitable for domestication. The addition of animal protein, manure and haulage to the Oaxacan economy would have allowed these people a wholehearted commitment to the sedentary farming life at a much earlier stage in their development.

Centuries later, the Iroquois Indians of North America, themselves agriculturalists working in the tradition that had first been established by early farmers in the Oaxaca valley, described the beans, squashes and maize that they grew as "three sisters, those on whom our life depends''. And although maize was certainly the staple crop in Central America, the other plants were no less crucial to their existence. Despite the fact that maize is a plant high in starch, it is relatively low in the total protein it provides; while it contains essential niacin, this occurs naturally in its bound form niacytin (early farmers learnt to add a pinch of wood ash to their corn dishes to aid the release of the niacin). Crucially, moreover, maize lacks completely certain amino acids that are vital to human life. These amino acids are found in beans and squashes, along with the extra protein that is needed.

Maize, beans and squashes provided the earliest Oaxacan farmers with most of the nutrients that they required, and it was as a sisterhood that these three crops were cultivated. Prodding the earth at regular intervals with small wooden digging sticks to make holes, the Oaxacans devised a method of planting the seeds of the three plants at different times of the year, so that all of them would come to fruition simultaneously. The workers then toiled day in, day out beneath the blazing sun, tending the young seedlings individually to coax each into abundant growth. The beanstalks were carefully trained to grow up round the sturdy stems of the maize for support, with the squashes hugging the ground in-between. Since this system crowded the plants into tightly packed ranks, leaving very little surrounding space, it was particularly important that any competition in the crop fields from encroaching weeds for the soil's precious mineral nutrients, and for the

light, should be fiercely resisted, and so the Oaxacan farmers plied their wooden hoes busily throughout the growing process to keep them at bay.

While the clearing of land, tilling of soil, sowing, constant weeding and harvesting involved in agriculture of this sort demanded an enormous investment of labour, the return in terms of produce more than made up for it. Central American communities, like other emerging agricultural societies throughout the world, found themselves able to support increasingly large populations, including many persons not directly involved in crop cultivation. As sedentary communities, moreover, they began to accumulate a variety of possessions — not only reserves of food but objects for which such reserves could be traded. These began with valuable tools and ornaments in numbers that had been impractical in nomadic times; eventually, the more successful farmers came to own small parcels of land, buildings and precious metals. Differences in wealth between individuals and families gradually began to emerge, and with them came the evolution of landowning and labouring classes.

Small villages grew into great cities as the prosperity agriculture brought generated more work in nonagricultural occupations. Craftsmen were engaged in manufacturing farming tools and weapons, and in creating ornamental artefacts for the rich. Engineers and builders were employed in building palaces and granaries, and staff in running them. Consolidating their power, rulers gathered bodyguards and armies about themselves as protection against resentful inferiors and rival magnates, and established bureaucracies to manage the institutions of state that kept them in place.

Powerful and complex as such new societies became, however, they were vulnerable in a way that hunter-gatherer groups, with their low population densities and generalist survival strategies, responding instantly and easily to the effects of local environmental changes, never had been. The greatest of the Central American empires, no less than the humblest village, was built upon maize, beans and squashes — and if drought, flood or any other natural disaster should cause the crops to fail, the edifice would crumble. This dependence was recognized: according to the Spanish priest Bernardino de Sahagún, writing during the 16th century in the early decades of Spain's colonization of Mexico, an Aztec of the day described maize in religious terms. It was "our flesh, our bones," he said. "I honour it. I desire it. I venerate it, esteem it. It is our sustenance."

Similar attitudes prevail among the Hopi Indians of northeast Arizona, in the U.S.'s arid southwest, many of whom still manage to wrest a living from the rocky, unyielding soils around their pueblos, or villages of mud huts, cultivating maize and other crops according to the methods developed thousands of years ago by the first American farmers. For the Hopis, nothing is of greater importance than the successful outcome of the maize crop, without which the life of the community simply could not continue. The maize plant itself — with its upright posture and its torsolike cob, with pollen-bearing tassels hanging down like tresses of hair — is seen as resembling the human form. Each stage of the growing process — sowing, sprouting, maturing and harvesting — is attended by ancient prayers and rituals, by communal dances and songs, all intended to encourage the maize as it struggles to grow. Each August the men of the pueblo gather to perform a snake dance, holding live rattlesnakes in their mouths, hoping that these reptiles will carry to the rain gods their pleas for showers to nurture the thirsty roots of the growing crops.

Just as Hopi rituals attend the annual progress of the maize crop, so the maize plays a leading role in all the myriad rituals surrounding the life cycles of men and women.

When a baby is born, for instance, a "corn mother" – a cob with four kernels grouped in a cluster at its head – is placed beside the infant for its first 20 days, during which time it is kept in a darkened room, its body deemed to be still under the protection of its universal parents, the earth and sun. At regular intervals throughout this period the baby is washed and then rubbed with corn meal. A line painted on the walls and ceiling of the room in corn meal completes the child's protection: such lines are seen as barriers to evil spirits, and are also used to block roads and trails. Each new day is saluted with corn meal, scattered before the rising sun. Used at every religious ceremony, corn meal's force as an offering stems from its identification with the human body: as the Hopis sprinkle its seeds during prayer, symbolically they are offering up a little of their own flesh.

The Mesopotamian goddess Inanna, shown above between symbols of fertility on a stone bowl of around 2500 BC, was one of several goddesses around whom legends were spun recounting the cycle of barren and regenerative seasons. Descending to the underworld to seek her dead husband, Inanna was stripped of her powers, but was resurrected after being sprinkled with the water of life. The Egyptian goddess Isis, shown suckling her child in a faïence statue from 1400 BC *(opposite)*, was married to Osiris, god of fertility and death; when Osiris was killed by the god of drought and his body scattered in pieces across the land, Isis restored him to life and they had a child, Horus. In Greek legend, the mourning of the goddess Demeter for her daughter Persephone, who was carried off to the underworld, caused the crops to die; the two were reunited by Zeus, but Persephone was bound to return to the underworld for a part of each year.

The Hopis have woven such a powerful web of emotion and religious devotion around their main source of sustenance precisely because they are uncomfortably aware, as were their predecessors, that this source may one day fail them, leaving them with nothing at all. Though relatively infrequent, disastrous crop failures have devastating consequences for farming communities, and are remembered for many generations afterwards. And if famine has proved an inseparable companion to agriculture since the first farmers tilled the soil, this has in large part been because the demands made upon the earth by agriculture have at times simply been greater than the earth could bear. Cultivated year after year, fields become exhausted, drained by the growing crops of all the nutrients and essential minerals they contain. Such land needs to be well and regularly rested. Farming communities discovered this problem early on in agriculture's history, but the unwary could still be brought to grief.

Even when apparently successful, agriculture still took its toll upon the earth. The clearance of forest for cultivation removed the extensive root systems that had locked soil firmly in place; once the trees were gone, the topsoil might easily be washed away by the first heavy rainfall. Trees are also far more efficient than other forms of vegetation in helping to dissipate moisture from the earth by evaporation. When they are cleared, the ground beneath loses this means of releasing its moisture, and apparently dry woodland might turn out, after deforestation, to be too boggy for cultivation. It was this process that led to the development of upland peat bogs in some parts of Europe, such as the Dartmoor region of southern Britain.

Another problem encountered by early farmers was salification. In the Near East in particular, where river water was used to irrigate drier areas through systems of canals and ditches, the salt and other minerals carried along in the water, and normally destined to accumulate in the sea, built up in the soil instead. When the water evaporated it left its salt deposits behind, to build up year after year. This process of salification significantly reduced the fertility of the soil; once salt levels of more than 1 per cent had been reached, the land became barren.

Other problems stemmed from the sedentary existence that was dictated by agriculture. Settled communities, packed together in villages or cities, in close proximity

to their livestock and indeed to their own waste products, ran a risk of disease that was never imagined by earlier hunter-gatherers, who were protected by their low population density and their nomadic ways. Great pestilences periodically swept the world's agricultural areas, bringing rapid and horrendous loss of life. Always a very demanding system of land use, in both labour and environmental cost, agriculture could at times prove disastrous for its practitioners.

Some 4,000 years ago, on the central Asian steppe between the Caspian Sea and the northern edges of China, peoples who had for generations been reaping sparse harvests from arid and unrewarding soils began to cut their losses. They abandoned the agricultural life altogether. Since their oxen, their horses and their flocks of sheep and goats had thrived on the limitless pasture of the steppe, it was naturally to this resource that they now turned.

The new pastoralists could not simply abandon their fields and continue grazing their old pastures, however. Because everything now depended upon their flocks, they had to increase their numbers and the size of the individual animals in order to enhance their productivity as a food resource, and this meant finding the best possible grazing and watering land for them all year round. Localized grazing, which was seasonally fluctuating and quickly exhausted by large flocks – could never meet the new standards. The pastoralists therefore had to move their flocks on from place to place, to wherever the best pasturing was to be found. Thousands of years after their ancestors had given up the wandering life of the hunter-gatherer, the peoples of the steppe found themselves nomads again.

If economic considerations made the new nomadism desirable, a number of other factors made it more practicable than it might have been in earlier times. Perhaps most vital of all was increased mobility. Over the centuries since its first domestication, the horse had been tamed to a still higher degree, and was now thoroughly broken to the pack and the saddle, dramatically extending the scope for human travel across the endless steppe. Travel was made even easier by the advent of the wheel, introduced from the agricultural Near East. While still having to shed many of the possessions that had surrounded them in their sedentary days, the new nomads could use carts hauled by oxen or horses to carry heavy loads never dreamt of by the hunter-gatherer bands of earlier ages.

Within the human body itself, meanwhile, a more intimate, physiological development had slowly, imperceptibly, been taking place. The enzymes that were secreted in babies' stomachs to permit the digestion of their mothers' milk, previously lost after infancy, were beginning to endure throughout adulthood. The impact of this change was revolutionary, for it added a second, and well-nigh inexhaustible source of rich animal protein to the pastoralists' diet. Where animals had once provided humans with nourishment in the single form of meat – and were productive only when killed – goats, sheep and mares were now twofold nutritional resources, supplying not only meat but also more long-term nourishment in the form of daily milk. Though it was often drunk in its natural form, a range of dairy products were soon developed to make milk more readily storable and portable: yoghurt, cheese and butter. In addition, dehydrated milk was prepared by heating milk until the solids separated and then drying these solids in the sun.

Mare's milk was especially prized by the pastoralists. Oxen were kept in small numbers for haulage but cattle, except in Africa, generally

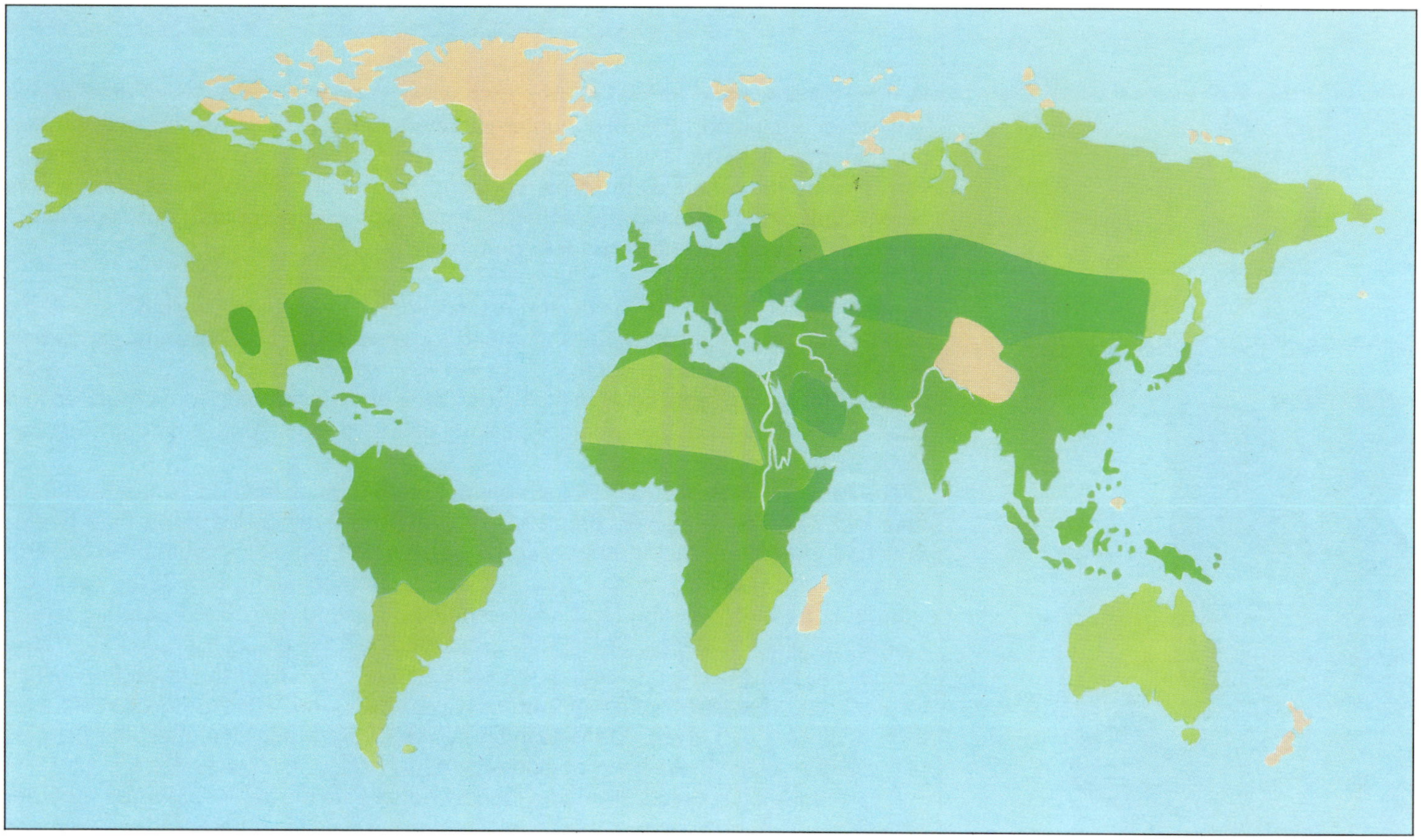

This map shows the approximate regions of the world that were occupied by hunter-gatherers *(shaded light green)*, by agriculturalists *(shaded mid-green)* and by pastoralist herders *(shaded dark green)* in the year AD 1. The areas coloured beige were as yet unclaimed by humans. Relative to the time span of human inhabitation of the earth, the encroachment by farmers and pastoralists over less than 8,000 years on territory first occupied by hunter-gatherers was spectacularly rapid. This expropriation of land by late-developing cultures was to be continued. City dwellers who depend on others for their food now comprise more than 40 per cent of the human population, while hunter-gatherers have dwindled to less than 0.001 per cent.

proved ill-adapted to the nomadic life style, consuming large amounts of water and high quality grass in return for milk that was thinner and less rich in vitamins than that of the mare. Mare's milk could also be fermented to make an alcoholic drink, which later Mongol nomads would call kumiss.

Milking was not the only means the pastoralists had of drawing nourishment from living animals: often it was tapped more directly in the form of blood, which could be used as an ingredient in cooked food or simply drunk neat. As long as only small amounts of blood were taken at a time, and the animals were bled in rotation to allow each one a few days to recover before its blood was taken again, the practice could continue indefinitely without any adverse effects on the pastoralists' livestock.

The pastoralists now had in their flocks and herds a food supply that was adequate to all their nutritional needs. As long as their livestock could be fed, they themselves need never go hungry. So, travelling for much of the year in small, family groups — for larger social gatherings would put overwhelming pressure on an area's pasture — the nomads would make their way across the steppe, moving on at regular intervals to new, unexploited pastures in endless pursuit of good grazing ground. Their unhurried flocks feeding as they went, they covered no more than a few kilometres a day, but there was little need for haste. There was usually, however, a destination, and a timetable of sorts. While a few of these groups wandered the countryside without any definite itinerary, and were permanently on the move, more often the nomadic pastoralists followed well-trodden routes, moving each year between their winter and summer grazing areas, often many hundreds of kilometres apart. Sometimes the migration would be from an area of low ground – mild and grassy in winter, but hot and sparsely pastured in summer – to an upland region that remained relatively cool, moist and green in summer, though it might be impossibly harsh in winter. In other cases the move would be a straightforward journey from north to south, taking advantage of changing climatic zones.

The nomads' flocks were far more than just a source of food. Their hides were used for making harnesses, their bones for tools. Their dung provided the perfect fuel out on the steppe where wood was very scarce. Wool was spun into cloth for garments, and for the elaborately woven rugs used as portable flooring for the nomads' tents. Wool was also used to make felt for the tents themselves. Laid out in great quantities on the ground, the wool was painstakingly flattened and rolled into tight cylindrical shapes until the matted fibres cohered as a single sheet. These sheets, waterproofed with animal grease, were then stretched upon frameworks of wooden poles, in winter often several layers deep, and held down by weighted horsehair ropes.

**By the time this ivory plaque of a cow suckling its calf was carved in Syria in the eighth century BC, milk was sustaining humans as well as the offspring of wild animals. Both hunters and farmers contributed to the domestication of animals by bringing home the young of prey they had killed; by keeping captured animals for eating at a later time and meanwhile feeding them on surplus crops; and by gradually exerting an increasing degree of control over the migrating herds they followed. By 4000 BC the basic herd animals – sheep, goats, cattle, pigs – were serving humans in ways that were unknown to earlier hunter-gatherers: they provided milk, wool and power for drawing ploughs, and their possession was a new indicator of wealth.**

Although in principle self-sufficient, the nomads' existence was often insecure. The lands they had abandoned as farmers had been marginal, but later agriculturalists often succeeded in bringing them under cultivation by using systematic irrigation, and the nomads found themselves pushed deeper into the barren interior of the steppe. Even good pasture was subject to occasional drought, while a single late frost, destroying the young shoots of spring, could blight a season's grass over vast areas. Like the agriculturalists, moreover, the nomads contributed unwittingly to their own problems, their very success sometimes leading to setbacks. If their flocks grew too large they placed a heavy burden on the available pasture. Persistent overgrazing could lay bare the lushest grassland, and destroy what trees and shrubs there were on the steppe. Without such cover, nothing kept the topsoil from being carried away by wind and rain, after which the land became useless desert.

The nomads seldom enjoyed the sort of surpluses that the agriculturalists did, and often suffered severe shortfalls. They were frequently anxious to trade their supplies

of livestock, felt, leather goods and woollen carpets and clothing for the produce of the settled agriculturalists near their migration routes — but the farmers, who generally looked upon their wandering cousins as irresponsible and untrustworthy vagrants, were often rather less eager to do business. The nomads therefore sometimes resorted to war to get what they required, making mounted attacks on village settlements in order to loot the contents of granaries and storehouses and seize the villagers' livestock. Such were the nomads' easy mobility, their consummate horsemanship and their skill in shooting their short bows from the backs of their galloping horses that they were usually victorious.

Periodically, some of these nomad bands would push outwards from the interior of central Asia, massing as they went in loose confederacies to seek both grazing and conquest. Their descendants would bring down empires in China, and in western Europe would destroy the might of classical Rome. But usually these nomad alliances were short-lived and broke down quickly once victory had been won — their social order was by definition a shifting, temporary thing. Outnumbered by their subjects, some conquering nomads adopted settled ways and were absorbed into stable agricultural communities; others quickly quarrelled with their fellow wanderers and parted, only to range the steppe once more.

In the Zagros Mountains of southern Iran, to the south of the Caspian Sea, a people known as the Qashqai have kept the nomadic ways alive down into modern times. Though cultivated fields and the factories and built-up areas of the city of Shiraz now block their ancestral migration routes, the Qashqai still manage to make the round trip between their summer and winter quarters each year, some clans covering around 1,000 kilometres with their flocks of sheep and goats. The tribe's hereditary nobility now own tracts of agricultural land in each area — apple orchards and vegetable fields in the north, citrus and date orchards and fields of barley and wheat in the south — and travel between them by lorry and jeep. But most ordinary Qashqai are still dependent upon their flocks for sustenance and make the annual migration in the traditional way, driving their flocks across rugged terrain on horseback and on foot, and living all year round in tents made from cloth of woven goat hair.

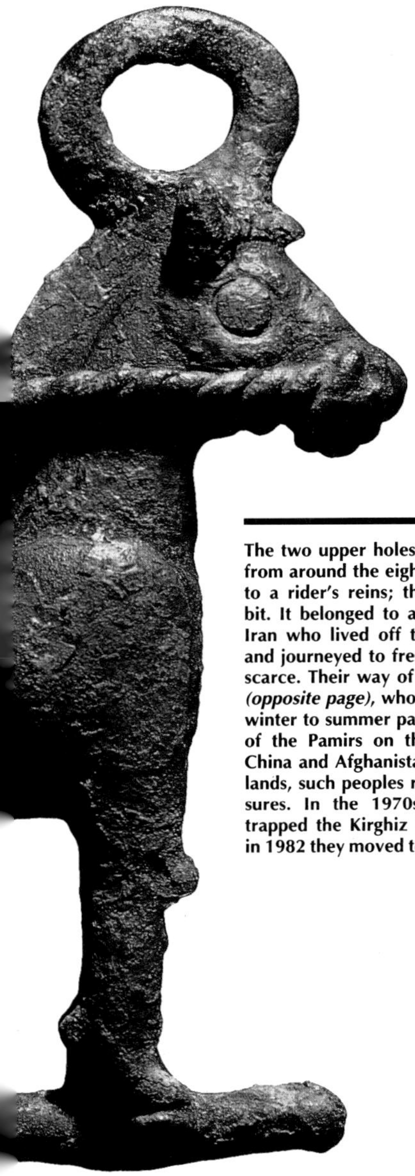

**The two upper holes in this bronze cheekpiece dating from around the eighth century BC are for attachment to a rider's reins; the central hole is for the horse's bit. It belonged to a seminomadic people of western Iran who lived off the meat and milk of their herds and journeyed to fresh pastures when grazing became scarce. Their way of life is perpetuated by the Kirghiz *(opposite page)*, whose passage with their camels from winter to summer pastures is framed by the mountains of the Pamirs on the borders of the Soviet Union, China and Afghanistan. Though able to survive in arid lands, such peoples remain vulnerable to outside pressures. In the 1970s, for example, border disputes trapped the Kirghiz in the mountains of Afghanistan; in 1982 they moved to a new life in eastern Turkey.**

In addition to an assortment of horses, donkeys and camels, the average Qashqai family also owns a flock of around 40 sheep and 30 goats, from which it gets meat, milk and dairy products, as well as wool and hair for making into cloth. These animals are common to most nomadic pastoralist peoples; camels are generally herded only in the driest regions, and cattle where there is plentiful rain and good supplies of grass. The goats provide a vitally important safety net: these hardy creatures are capable of surviving the kind of conditions that would kill off most other animals, and are the last to cease producing milk when the land is stricken by drought. All animals kept by pastoralists tend to be very lean and spare, but comprise nevertheless their owners' chief form of wealth. (The word pecuniary, meaning to do with money, derives from an ancient Sanskrit word for cattle through the old Latin *pecu*, meaning cattle as wealth or property.)

The Qashqai spend the winter months in the green hills of Khonj above the Persian Gulf, where the men tend their flocks and hunt ibexes and other game for sport — like their central Asian forebears, they still shoot their prey from horseback with fast and deadly accuracy. The women look after their families, spin wool and weave carpets.

Each year in late March the community dismantles its tents, packs up its belongings and loads everything on to the backs of horses, camels and donkeys to decamp as one, driving its flocks northwards, raising as it goes a dense pall of thick dust that looks from a distance like the advancing smoke of a great fire. At the end of each day's journey the men raise the roofs of their family tents on long poles and batten down the sides with wooden pins; the women and children pound stakes into the hard earth; rugs are shaken out and spread on the ground, the weary pack animals are watered and fed, and fires for cooking are lit.

A present-day member of the Qashqai remembers in his youth seeing a great throng of 5,000 families crowding through mountain passes and crossing fords, accompanied by perhaps as many as half a million sheep and goats and some 70,000 horses and donkeys. Their numbers today are much diminished. Many of their traditional pastures were nationalized in the 1950s and the land redistributed to settled farmers; new roads were cut across their grazing lands, destroying the grass; and in tent schools erected in their summer and winter camps the Qashqai children were taught to read and write, enabling them to get jobs in towns. But for those remaining few who have not forsaken the old ways, the steady rhythm of seasonal migration still continues, and for almost two months the Qashqai continue their journey, piloting their hardy animals across apparently trackless wastes, steering them through fast-flowing rivers and negotiating treacherous crags and rock faces, until they arrive at their summer quarters. A few short months later, with the arrival of autumn, it is time to begin the return trip once more.

The Qashqai of Iran, the Hopis of Arizona and the !Kung of the Kalahari are, sadly, rare survivors. In most parts of the world the techniques of survival practised by the earliest human inhabitants – techniques that evolved primarily in response to local climate and conditions – have long since been replaced by strategies based on the deliberate adaptation of nature to meet human needs. Many hunter-gatherer peoples in particular, over the course of the last 500 years or so, have been driven to extinction by the increasingly sophisticated weaponry of invading armies or by the expropriation of their foraging grounds. Those groups who have succeeded in escaping this fate live mostly in desert, rainforest or arctic regions where agriculture is impracticable or uneconomic, or have been reduced to dependency on reservations. They have suffered, too, from the arrogant misconceptions of succeeding cultures, epitomized by the opinion expressed by the English political philosopher Thomas Hobbes in 1651: "No arts, no letters, no society, and, which is worst of all, continual fear and danger of violent death, and the life of man solitary, poor, nasty, brutish and short." Only one of these generalizations – the absence of letters, or a system of writing – in fact has any accordance with reality.

Because of the endurance of such deep-set prejudices, the fact that the !Kung and other peoples like them should choose to continue with their ancient ways seems surprising to modern city dwellers. Doubtless, the life of those who choose to live in the cities is as perplexing to them. Arduous and materially impoverished as their spartan, nomadic life styles may appear, they endure precisely because they have succeeded in meeting the needs of their practitioners over many hundreds of years, and because, despite disruption by modern cultures and governments, they continue to meet those needs today. Their persistence represents not a failure of development but the triumph of economic systems so perfectly matched to their environments that no further development was required.

# SANCTUARIES OF THE SPIRIT

The earliest humans, watching lightning flare in the sky or angry, storm-driven waves thunder against a sheer cliff face, could not doubt that the earth was spirit as well as matter. And the spirit was made manifest not only in the awesome spectacles of weather but in the daily round of existence: at times the earth provided abundantly for all human needs, at others its bounty was withheld. Clearly the natural world was instinct with energies far beyond human ken.

In all parts of the world, certain places in which the immanence of the earth's spirit was most acutely felt were accorded a special reverence. Prominent among these sacred sites were particular mountains, rivers, waterfalls, caves and groves of trees. Many were distinguished by the extreme beauty of their natural formation or setting, as shown on the following pages. From the tumbling and turbulent waters of the 226-metre-high Kaieteur Falls in Guyana, shown on the right, for instance, spray continually dances in clouds of ever-changing hues. When a drought reduces the volume of water passing over the falls, a curious stone shape can be seen at the base, which is believed by local people to be the petrified canoe of a king who offered himself to the spirit of the falls.

The great web of legends and myths that accumulated around all such sites gave narrative form to their inherent natural forces. Rites of worship and religious cults helped to protect them from desecration, a crime for which the penalties could be severe. During the first century AD, for example, a Roman general who cut some timber for shipbuilding from a grove dedicated to the god Asklepios on the island of Kos was summarily executed in the grove itself. Wherever the bounds of the sacred were transgressed – whether by an army of occupation or by a company seeking profit from mineral deposits – the pain of violation was felt by all those who deemed the site holy.

Dyed red by the setting sun, the furrowed sandstone cliffs of Ayers Rock rise above the barren plains of Australia's desert interior like the humped back of some stray subterranean creature. Eight kilometres in circumference and 348 metres high, the rock was named after the premier of South Australia in around 1873 by an explorer ignorant of its sacred status in the mythology of the Aborigines, to whom it was known as Uluru. The origins of the caves and potholes that puncture the rock's smooth surface are accounted for in the legends of the Aborigines concerning their ancestors, and Ayers Rock has now become a place of secular pilgrimage for thousands of tourists every year. In other parts of the world also, certain mountains and other high places have remained sacred to successive cultures. Mount Ararat in eastern Turkey, for example, was in local Armenian lore associated with a great flood that engulfed the world before its identification in the book of Genesis as the place where Noah's Ark came to rest. In Japan the serene, symmetrical Mount Fuji was held sacred by the aboriginal Ainu people centuries before a shrine was built on its summit by adherents of the Japanese Shinto religion.

An artery of molten silver in pale morning light, the River Vilcanota meanders beneficently through the Urubamba Valley in Peru, the heartland of the Inca empire from the 12th to the 15th centuries. Temples built in the valley testify to the reverence accorded by the Incas to the river's life-giving waters, which nourished the agriculture on which their prosperity was based. The Vilcanota was to the Incas as the Nile was to the ancient Egyptians, who 4,000 years ago addressed a hymn to the spirit of the river "who issues forth from the earth, who comes to bring life to the people of Egypt". In India the River Ganges still attracts such worship: in 1986 four million pilgrims attended a festival at a town in the river's foothills. The Ganges for Hindus is Ganga, the goddess of purification. A verse in the Sanskrit epic the *Mahabharata* declares that "If only the bone of a person should touch the water of the Ganges, that person shall dwell honoured in heaven."

A circle of standing stones echoes the rounded shape of the mound behind it. This circle in Cumbria in northern England is one of many that were erected in northern Europe — more than 900 in Britain alone — by farming communities over a period of 2,500 years from the fifth millennium BC. The avenues framed by 3,000 standing stones at Carnac in Brittany and the earliest megalithic structures at Stonehenge in southern England were also built during this period. The function of the circles was probably in part astronomical, the siting and arrangement of the stones being related to movements of the sun, moon, stars and planets. But whatever their precise purpose may have been, the space which they encompassed was hallowed ground. By the pattern of the stones humans sought to relate themselves to the more mysterious order of their natural environment: their erection was an act of homage.

# THE MEDITERRANEAN WORLD

2

Sometime around 700 years before the birth of Christ, two farming brothers in Boeotia, in what is now mainland Greece, got into a disagreement with each other. Both farmed their own smallholdings, but while one of them was a hard-working and prosperous farmer, the other, Perses, was a lazy and careless waster who eventually had to fall back on his more careful brother for a loan. Perses got more than he bargained for. Rather than simply hand over money or grain, or lend him a plough, his brother decided to give him something far more valuable. He wrote for Perses an epic poem that contained in around 820 lines all the advice a farmer could possibly need in order to prosper: the specific age of the oxen he should buy for pulling a plough, the precise dates on which various crops should be planted and harvested, and the most effective way to store surplus crops to fend off possible future hunger. Perses learnt also how to avoid offending the array of gods upon whom much of his prosperity would depend. Urinating facing the sun, for example, was bound to upset Helios, the sun-god. Armed with his brother's wisdom, Perses could not fail to do well for himself, although even then he could not expect it all to be easy – the key to survival, the poem warned, was hard work.

Perses' brother, the author of this hymn to prudent husbandry, was the farmer-poet Hesiod, of whom it was said that he learnt the art of poetry from the Muses themselves while he tended his sheep in the hills. It was also reported that after he was killed – as punishment for fathering an illegitimate child – his body was tossed into the sea, but was borne back to shore by dolphins.

Much of Hesiod's poem for his brother, entitled *Works and Days*, portrays a landscape which is not unlike that of present-day Greece, where farming is a constant struggle against the handicaps of poor soil, barren slopes and scorching, rainless summers. Never far away from Hesiod's mind as he gave his advice was the threat of hunger, even of actual starvation. And if the life of an individual was precarious, so too was the very survival of whole early populations. For all the material wealth and cultural achievements of the successive civilizations which inhabited the Mediterranean region, famine remained a constant and ominous spectre. The valley bottoms were filled with the golden wheat of scattered farms, and the foothills would in time become covered with the silver-green of olive trees and the darker green of vines, but that was still not enough. Beyond the cultivated lower slopes rose white limestone mountains, some of them as much as 3,000 metres high, gouged with deep ravines and gorges. In places these were forested with pine and oak trees, but more often they were covered with maquis – a dense scrub made up of juniper, shrubby oak and wild olive. A drab brown in the summer and a vibrant blaze of colour in the spring, this land was virtually useless for agriculture throughout the year, good at best for rough grazing. Elsewhere, many of the streams ran dry in the summer, while

**This Greek terracotta tableau showing a farmer trudging along behind his ox-drawn plough dates from the seventh century BC, some 400 years after iron was first used in the Mediterranean lands to make ploughshares, or blades. The eighth-century-BC poet Hesiod recommended that the plough beam and the stock – the part held by the farmer – be made of oak, and that the plough should be yoked to two nine-year-old oxen. Where the soil was heavy, a team of as many as eight animals was required. Hesiod also suggested that the plough be guided by a responsible 40-year-old labourer "who'll keep his mind on work, not look around for friends, the way a young man would."**

during the winter months driving rain washed away fertile soil from the hills, and swollen rivers carried precious minerals from the earth uselessly out to sea.

The achievements of the successive peoples who inhabited the Mediterranean region derived above all else from their delicate relationship with this obstinate and often inhospitable landscape. The natural world shaped their development: it was the abundant fertility of the plains of Mesopotamia that allowed the Sumerians to build the world's first cities; and without the mountain ranges that restricted the size of Greek city-states to manageable proportions, and the warm sunlight that bathed their open-air meeting places, the Greeks might not have been capable of developing the prototype structures of political democracy.

This shaping was not just a one-way process, for, in their struggle to overcome the limitations of their climate and landscape, the peoples of the Mediterranean would exert strong influences of their own. At first, when the population was relatively low, farmers and pastoralists were able to satisfy their needs simply by living in cooperation with the seasonal rhythms of the land they lived off. But as their numbers inexorably grew, and food supply became a pressing problem, they resorted to increasingly ingenious ways of working the soil and making the most profitable use of their limited resources. In so doing they not only altered the face of their homeland, but also discovered that not just rivers and crops but whole civilizations too might be subject to rhythms of growth and decline.

**Man-made steps cut into a hillside on the Greek island of Corfu permit the steep, rugged terrain to be put to the maximum productive use. Throughout the Mediterranean such hills were once covered by forests; after the trees had been felled for fuel or construction, terraces reduced soil erosion and retained moisture, and from around 3000 BC were planted with olive trees. The Greek amphora *(inset)*, dating from approximately 520 BC, is decorated with a harvest scene: the olives are knocked down with long poles, then gathered into a basket. The jar itself was probably used to store olive oil, which was used in cooking, for lighting and, in the absence of soap, for cleaning the body.**

Around 10,000 years ago, when nomadic peoples first began to gather together in settled communities in the Fertile Crescent around Mesopotamia for the common purposes of planting crops and stockbreeding, the environment in which they dwelt itself provided the solution to most of their practical needs. Every spring the snows of the Zagros Mountains and the Armenian Highland melted, swelling the Tigris and Euphrates rivers to coffee-coloured torrents that washed down over the flatlands. The flood waters carried with them a mineral-rich alluvial sediment of fine soil particles mixed with clay and gravel. This material, known as silt, was the Mesopotamian farmer's main source of fertilizer for the soil on which the richness of his crops depended. The flood water served to irrigate the land. In addition, the rivers provided a simple means of communication and transport, so that surplus produce could be easily exchanged with other communities.

Faced with the natural rhythms of flooding, seed time and harvest time, the Sumerian people of Mesopotamia assumed that they participated in a divine order, and that their society had to reflect the orderly ways of the gods. Their religion centred on the worship of an earth mother, while the embodiment of male fertility was the bull, and the seasonal decline and renewal of the soil was marked by ritual bull sacrifices. Other early agricultural societies followed similar fertility cultures, some involving human sacrifice, usually of a king who had ruled for a year before being offered up to the goddess, his blood representing the fertilization of the soil.

The farmers prospered, and by 3300 BC they had established the world's first urban centre at Sumer, on the plain between the Tigris and Euphrates rivers. Shortly afterwards the city of Memphis was built in Egypt, on the banks of the Nile. As would become the pattern throughout the Mediterranean, these cities were founded on agricultural prosperity: only where the land provided enough food for people not directly involved in its production were trade, specialist crafts and an urban way of life able to develop. But while the new cities themselves accrued great wealth, especially for the emergent ruling classes who controlled the labour of the majority, in the process they changed the whole relationship of their inhabitants to the land from which they had come. The fertile countryside was now under the political control of the cities, and subject to new and exacting demands.

A Sumerian poem known as the *Epic of Gilgamesh* recounts a legend which encapsulates this change. The hero Gilgamesh, ruler of the city of Uruk, determines that he will challenge the god who protects the forests: "I will conquer him in his cedar wood and show the strength of the sons of Uruk, all the world shall know of it. I am committed to this enterprise: to climb the mountain, to cut down the cedar, and leave behind me an enduring name." The Sumerian cities did indeed require vast amounts of timber for house construction, pottery firing and the smelting of metal ores. But Gilgamesh's successful completion of his task invoked the wrath of the sky god Enlil, who threatened to wreak a terrible revenge on the people of Uruk: "From henceforth may the fire be on your faces, may it eat the bread you eat, may it drink where you drink." In practical terms the consequences of the clearance of forests around the headwaters of the Tigris and Euphrates rivers were no less severe: the soil on the hillsides that had previously been anchored in place by the trees was now washed down by the rivers, gradually clogging irrigation canals and turning the lowland plains into a barren, salty waste. Without the support of its abundant agriculture the Sumerian civilization soon weakened and lost its dominance.

**In a Roman stone relief, two labourers pound rods to assist their stamping feet in treading grapes to make the wine that the Romans, like the Greeks, often drank diluted in preference to water. The resultant mush was squeezed by a mechanical press, and the juice poured into resin-coated vats to ferment. The fourth-century-BC plate *(opposite page)* is decorated with images of a red mullet, a bass, a torpedo fish, a sargus (now extinct) and a cuttlefish. In the time of Homer in the ninth century BC, fish was regarded as a food of the poor; in later centuries, when wild game was depleted and arable land was used for crops rather than stock rearing, fish was for all classes a valued supplement to their mainly cereal and vegetable diet.**

Meanwhile the people of Egypt's new cities continued to flourish along the narrow belt of fertile land on either side of the Nile, not least because they had found a rich source of essential timber that did not deplete their own limited supplies. Paintings found in tombs at Thebes record the visits to Egypt of traders who came from the three-peaked and wooded island of Crete, which by around 2000 BC had become one of the Mediterranean's most powerful states. In the Near East, where cedar wood was in short supply, its use was mainly reserved for palaces and temples – on Crete, however, it was so abundant as to be used to make everyday objects, and the island was also richly endowed with oak and pine.

Trade brought great wealth to the Minoan people of Crete, and their rulers built themselves vast palaces at Phaistos, Mallia, Zakro and, pre-eminently, Knossos. Each of the palaces became the centre of its own local hierarchy of bureaucrats, craftsmen, traders, priests and priestesses, and agricultural workers, and provided a degree of centralized authority over the agriculture and trade in its vicinity. At the same time the religious practices of earlier cultures continued to be echoed in Minoan rituals that were centred on the bull: the god-king, known as Minos, was believed to be the human incarnation of the bull spirit, and on ceremonial occasions he would wear a bull mask to perform fertility rites.

As the palace economies continued to thrive, wood came to be overtaken in importance by other commodities. Hilly and sun-drenched, Crete was also blessed with sufficient rainfall to nourish diverse forms of agriculture. Among these were sheep farming: the palace at Knossos alone controlled flocks which numbered more than 100,000. Although the island lacked the plains that were necessary for any large-scale crop farming, fields and orchards yielded beans, lentils, peas and lettuces, and fruits such as figs and plums. The sea provided plentiful octopus, squid and other fish. Resin obtained from the cypress pine was used for incense and also as a varnish for preserving leather and metal. And as well as storage facilities for stockpiling surplus grain – which guaranteed continued food distribution in times of poor harvest – the Minoan palaces were stocked with giant ceramic jars containing wine and olive oil, Crete's most important exports.

Wild vines grew in abundance on the hilly island, and the Minoans were quick to discover the pleasing qualities of the fermented juice of their grapes. Wine would become as common a beverage in Crete and throughout Greece as beer had once been in Mesopotamia, where a large proportion of the grain that the nation produced had been given over for fermentation into alcohol.

Olives, too, grew wild on Crete, and in time became even more important than the vine throughout the whole Mediterranean region – the Greek dramatist Sophocles would later describe the olive as the "sweet grey foster nurse" of his race. The olive is a notoriously slow-growing tree, which takes over 10 years from sapling to reach its mature state, and its cultivation required both forethought and the willingness to invest resources and toil in a product which had only long-term gain. But the olive was also renowned for its hardiness – it was able to survive droughts that killed off virtually all cereal crops – and the ability of its thin, shallow roots to extract sustenance from barren soil in which little else was capable of growing. By as early as the end of the third millennium BC the Minoans were cultivating the silvery-leaved trees in extensive plantations. The oil that was extracted from the fruit proved one of the most versatile and valuable products of the entire Mediterranean world – a fuel for lamps, a basis for perfume or cosmetics and, most importantly, a nutritious food that proved much more dependable than cereal crops.

But while the olive, once it had matured, went on renewing itself yearly, useful construction timber was altogether less inexhaustible. To provide necessary building wood and fuel for its potters and metalworkers, Knossos alone – increasing by 28 times its original size within 1,000 years – required extensive forests to be cut down. Gradually builders began to use less timber in house construction, and potters sought alternative fuels. Clearly there was no longer enough forest left on Crete even to supply the Minoans' own needs, let alone those of their trading partners. Even the vital Minoan trading fleet, lacking supplies of timber for repairs and the building of new vessels, began to deteriorate.

By 1450 BC the Mycenaean people living on the Greek mainland had not only profited from the Minoans' decline but had taken over the island of Crete itself, and they proceeded to establish a rich trading empire of their own around the Mediterranean. Timber again served as a vital mainstay, especially to provide a fuel for the Mycenaean bronze and pottery industries: the town of Messenia alone was home to at least 100 bronzesmiths, and pottery was produced on a large scale for mass consumption and an international market that spread as far as Italy, Cyprus and Palestine. But as agriculture boomed and the population rose, and as cities

**In a relief inlaid with shell dating from 2400 BC *(above)*, a ram is sacrificed in Mesopotamia, where the entrails of sheep were read by diviners who claimed the ability to foretell the future. The 13th-century-BC Mycenaean pottery vessel cast in the shape of a bull's head *(right)*, was used for offering libations: in the same way that animals were offered up to the gods, liquids such as wine or oil were poured through the bull's mouth on to the ground to propitiate the spirits of the earth or the underworld.**

expanded in size and new settlements were created, land and timber became more scarce. By 1300 BC woodcutters were being forced to venture into the very heart of the Greek peninsula in search of suitable trees to fell.

When the cutters had done their work, nothing was left to protect the earth from the heat of the summer sun or the driving rains of winter. Soil was washed away from the hills by streams that flooded, the slopes around them having lost their capacity to hold water. The plains flooded, too, and valuable crops were smothered by silt. Despite the construction of dykes and ditches to divert the water, many regions were beyond repair – rich, brown forest soil had been eroded down to the level of red subsoil or white limestone bedrock. In addition a period of drought caused the failure of successive harvests. The result was an agricultural crisis, compounded further by a decline in the essential Mycenaean trades of metallurgy and pottery, dependent for fuel on now-scarce wood. Whole towns that had been based upon these industries were probably abandoned. The population figures plunged. In the region of Arcadia, a mountainous inland area that still retained its woods, refugees from famine sated their hunger by feeding on acorns from its oak trees.

With the ending of the Mycenaean domination of Greece, the countryside itself fell into a period of decline. Sheep and goats that were pastured on marginal land in the foothills made their own contribution to the process, destroying young shoots that bound the earth together and wearing away terracing that had been built for the cultivation of olives. In Greek Asia Minor the effects were still apparent much later: the Greek geographer Strabo noted in the first century BC how so much land was being washed away by one particular river that local farmers had decided to sue it:

## RITES OF SACRIFICE

**The ritual killing of animals was considered by many early societies the most effective means of communication with their gods. By making a gift of something they valued, humans hoped in return to obtain divine favours – a fruitful harvest, perhaps, success in war or a cure for infertility. Through the art of haruspicy, in which a skilled priest discerned omens of future events in an animal's entrails, sacrifice also offered a means of interpreting the divine will. Such practices were common to many early cultures, as the examples shown here and overleaf confirm, and the role of sacrifice as a bond between worshippers and their deities persisted in many later-established religions, including Christianity.**

Marked for sacrifice by a ribbon, a boar in prime condition is led towards its death in a Roman bronze relief from around the first century AD. In Roman religion, the pig was sacred to the goddess Demeter.

"lawsuits are brought against the Maeander River for altering the boundaries," he reported, adding that when the river was convicted "the fines are paid from the tolls collected at the ferries." The bare bones appearance now characteristic of some regions of Greece became so familiar that later generations would assume it was the natural and immemorial face of the land.

Other regions of the Mediterranean were not so diminished. The Phoenicians, from the area that later came to be known as the Lebanon, established an extensive maritime empire based on timber and textiles, the latter often dyed with secretions obtained by boiling vast vats of small sea molluscs. (The name Phoenician itself probably derives from a Greek word meaning "purple dye", referring to the valuable colour that in many regions was reserved for the highest members of society.) Around 700 BC the Etruscans, who came from the western coast of Italy along the Tyrrhenian Sea, also joined the trading routes, capitalizing on their rich natural reserves of metal ores: their hills held deposits of copper, lead, tin and, most importantly, iron, which had become the most sought-after metal for fashioning into plough blades, tools and weapons. And gradually the people of Greece, too, learnt to survive and even to prosper on their altered landscape, bowing their heads to the soil and often looking back with longing to a golden age when, in the words of Hesiod, "ungrudgingly, the fertile land gave up her fruits unasked".

Hesiod attributed the decline of the Grecian soil to the gods' dissatisfaction with humankind. But Hesiod's gods were very different from those of earlier agricultural

**A 19th-century scroll painting shows Ainu families feasting before the staked-out body of a sacrificed bear. The Ainu – an aboriginal hunting-and-fishing people still surviving in small numbers in northern Japan – revered the bear as a sacred animal and communed with their god by eating its flesh and drinking its blood. The bear's soul, they believed, would intercede with the spirit world on their behalf. Captured bear cubs were raised in cages for about two years and then killed in an elaborate ritual. At the following feast, the bear's pelt with the head attached was included in the celebrations by being offered a stew made of its own flesh.**

communities. While the Sumerians, for example, believed that they worked the land in order to please their fertility deities, Hesiod clearly laboured only for himself, sometimes even in the face of obstacles that the gods had put in his way. Although still immortal and possessed of supernatural powers, Hesiod's gods were also distinguished by very human qualities: they were fickle, envious and easily hurt. These qualities were shared too by the gods of Homer, the Greek poet of Asia Minor who composed his epics the *Iliad* and the *Odyssey* in around the eighth century BC. They foreshadowed a new trend in Greek thought towards a rational, more human-centred explanation of the natural world.

In one significant way, however, the gods of the well-born Homer and of the farmer Hesiod were very different creatures. The former concerned themselves mainly with prowess on the battlefield – usually the preserve of the rich, who were the only ones able to afford the expensive horses necessary for heroic warfare – and were of little or no use to poor farmers. Hesiod's gods appreciated hard work on the land, not the idleness of the wealthy elite whose wellbeing depended on the labour of others. This contrast reflected the developing division that was opening between those who worked the land and those who did not.

According to Hesiod, the lords who acted as magistrates were crooked: they ground down their fellow men and did not fear the gods as they should do. The cause of Hesiod's own bitterness was partly personal – the judges had favoured his brother Perses when the two had contended the distribution of the land left to them when their father died. But the complaints of the grouchy farmer probably also reflected a more widespread resentment. Many of the Greek aristocrats were huge landowners with widely scattered estates that were worked either by tenant farmers, who were obliged to surrender a sixth of their annual harvest to the absentee landlords, or by peasants who possessed no land of their own. When Homer had his warrior-hero Achilles imagine the most wretched life he could, Achilles thought of just such an existence, as a poverty-stricken labourer.

Antagonism between rich and poor, mirrored in the division between city dwellers and those who lived in the countryside, attended all attempts by the Greeks over the ensuing several hundred years to make their land yield enough food to support their growing population. By 600 BC there had emerged a number of powerful city-states, each of which was divided from its neighbours by high mountain ranges, and their populations soon grew too large to be adequately fed and clothed from the meagre resources of their hinterlands. The problem became especially acute in Attica, a mountainous tract of just 2,500 square kilometres that jutted into the Aegean north-east of the Gulf of Corinth: this region, one of the driest in the whole of Greece, whose variable rainfall during the growing season frequently caused the crops to fail, had to support Athens, the largest city of all.

One ingenious solution was to reduce the levels of consumption by simply exporting the consumers. Most early cities on the heavily indented peninsula of Greece lay near the coast, within sight of the sparkling blue waters of the Aegean, and as food shortages became increasingly chronic, their rulers looked not inland but abroad. Thousands of citizens were dispatched in a great wave of colonization which saw cities established as far away as Spain to the west, North Africa to the south, and Turkey and the Black Sea to the east.

The colonists who were allotted their own plots of land around the new settlements were the lucky ones, for within Greece itself the situation was steadily worsening.

One adventurer, sailing from Corinth to the colony of Syracuse in southwest Italy, was reported to have become so hungry that, in desperation, he sold his promised holding for a honey cake. Because ownership of most of the good agricultural land was concentrated in the hands of a wealthy few, and because those few were increasingly planting valuable cash crops such as vines and olives rather than less profitable grain for food, poorer farmers were forced to make use of more marginal land with worse yields. There was little chance for them to improve their situation. They took out loans to keep their heads above water but, if their crops failed, they consequently found themselves in debt. The result often was that they were forced to serve their creditors as virtual slaves.

The condition of Athens became so desperate that in 594 BC the poet and lawmaker Solon was appointed to act as a mediator between the rich and the poor. To ease the situation Solon abolished all agricultural debts, allowed better-off farmers to become involved in the process of the city's government, and banned the export abroad of all produce from Attica except olive oil. Other states had different answers to the problem: in Sparta, for instance, the most valuable land was held by the state and farmed by a special slave class, the helots; in the Arcadian plain, every free citizen toiled with his own hands. But squabbles between rich and poor continued, and in some cities the discontent was exploited by ambitious individuals to provide them with an excuse to overthrow the government and establish their own tyrannies, which sometimes led to an improvement in the conditions of the poor.

The challenge of scarce resources was one of the factors that led inquiring Greek minds to investigate the laws of the natural world. During the fifth century BC the physician Hippocrates sought rational rather than divine or mystical explanations for the diseases he encountered; the historian Herodotus, whose interest in the customs and characteristics of foreign cultures was stimulated by his travels in many Mediterranean countries, came to believe that people's physical, mental and moral qualities were largely determined by their climate and environment. Such theories diminished the traditional influence of the immortal gods, and suggested that human problems could be solved only by human solutions.

By the middle of the fifth century BC the population of Attica numbered more than 200,000. Their diet was by no means extravagant. Bread was the staple, often dipped in wine, and this was supplemented by goat's cheese and fish, and vegetables such as beans and cabbage — most people tasted meat only on feast days. But because less than half their land was cultivable, they had become dependent on imports from grain-rich regions around the Black Sea and from Egypt, and the food supply had become a crucial political issue. Strict regulations stipulated that all voyages made by merchant ships financed by Athens must result in the import of food, and that no person living in Athenian territory could transport grain to any port but Piraeus, which served Athens. The penalty for breach of these laws was death.

For the Greek city-states the problems of food supply could only be solved by human intervention and the direct political management of natural resources. This brought with it new dangers, as the Athenians discovered: their dependence on supplies from abroad rendered them

**A fifth-century-BC bronze relief portrays the squatting figure of Pan, a fertility god of the fields and woods, playing upon pipes he has fashioned from a bed of reeds into which a fleeing nymph had changed herself. Roused from his midday slumber, Pan was lustful and unpredictable, liable to induce panic in both flocks and humans. His half-human, half-animal form personified the Greeks' ambivalent feelings towards the natural world: like themselves, nature could be both tranquil and unruly, and the border between what was civilized and what was bestial was never definite.**

more vulnerable to attack, and when their fleet was destroyed by their Spartan enemies in 404 BC citizens starved to death and Athens had no choice but to surrender its independence. But a pattern had been set, and the new overlords of the Mediterranean were to develop it even further.

In around 500 BC, when the Roman republic was founded, the Italian peninsula was inhabited by a number of independent peoples, and much of the fertile land had already been tamed and put to beneficial use. Forests had been cleared around villages to provide pasture for cattle, sheep and goats, or fields for growing grain and other crops. The Etruscan rulers of central Italy, the most powerful people in the region in the period before the rise of Rome, had divided their estates into small plots for the local population – who also benefited from Etruscan schemes of land reclamation and drainage, which enabled cultivation of areas that had previously been marshland or subject to damaging flooding. But the Romans had more far-reaching political ambitions than any of their neighbours, and in extending their territorial might they also set in motion a process that would remodel the landscape not just of Italy, but of the entire Mediterranean region.

The Roman republic at its inception controlled an area of around 1,000 square kilometres. By 260 BC this domain had increased to 25,000 square kilometres, and Rome had forged alliances with other Latin states occupying a further 110,000 square kilometres. At the end of the first century AD the fully fledged Roman empire extended from Spain in the west to Syria in the east, and from Britain in the north to the Sahara Desert in the south. As with other empires both before and after, once the process of territorial expansion had begun, it acquired a momentum of its own. In the case of Rome the impetus which drove the process was the need to feed and clothe an ever-expanding population, and to equip and pay for ever-larger armies – a need which could only be satisfied by gaining access to the natural resources of new provinces.

Within Italy the increasing pressure on agricultural land led to changes both in the traditional system of landownership and in the uses to which the soil was put. The citizens of the early republic were mostly subsistence farmers, smallholders providing for their families and trading their surplus produce. Conscripted to fight in an endless succession of foreign campaigns, many of these citizen-farmers were killed; others returned to find their farms overgrown. In due course their land was taken over by the government, who let it out to private individuals. Most of the best land was quickly snatched up by enterprising aristocrats, who amalgamated numbers of such smallholdings into large estates. Slaves brought to Italy from the colonies abroad provided cheap labour, and the absentee landlords soon discovered that cash crops represented a better investment than subsistence farming. As a result

PALUDES
VIA APPIA
COLONIA
AXVRNAS
2677

the production of oil and wine for export began to take precedence over the growing of corn for local consumption, and much land was given over to large-scale cattle and sheep ranching for meat, cheese, leather and wool.

An agricultural treatise written in the second century BC by Marcus Cato the Elder, who himself owned several estates, gave detailed instructions on how to establish a large mixed farm for growing grapes, olives, timber, fruit and cereals. The book included advice on the treatment of sick animals, and on how to maintain the fertility of the soil by using the dung of domesticated animals as a fertilizer. In such practices human ingenuity was being deliberately applied to the natural world with the aim of achieving maximum yields and profits. The new attitude was clearly expressed in a play by the Roman consul and writer Cicero, whose character Balbus the Stoic declares: "We enjoy the fruits of the plains and the mountains, the rivers and the lakes are ours, we sow corn, we plant trees, we fertilize the soil by irrigation, we confine the rivers and straighten or divert their courses. In fine, by means of our hands we essay to create a second world within the world of nature."

In the imposition of this "second world" the Romans made dramatic use of their engineering skills. Spectacular aqueducts were constructed for irrigation purposes and to bring supplies of fresh water into the very heart of cities. Between the fourth century BC and the early second century AD, nine aqueducts measuring a total distance of 425 kilometres were constructed around the city of Rome; the aqueduct at Nîmes in western Provence was 50 kilometres long, and was lined and roofed with stone slabs. To speed the transport of fresh produce and keep the outposts of empire in close touch with the centre of government in Rome itself, an elaborate network of roads was laid down across the landscape. Built wherever possible across level terrain, these paved highways followed the most direct routes to their destination, often striking directly through marshes or woodland, and were carried across rivers or deep ravines by soaring bridges and viaducts.

**A map prepared probably in the eighth century AD to illustrate a text by a Roman surveyor shows a colony founded in 329 BC in west central Italy after the conquest of a local people. A grid of lines intersecting at right angles divides the cultivable land into squares called *centuriae* – in theory, each held 100 heritable plots. The territory was marked out using a surveyor's cross, whose vertical alignment was established with plumb bobs *(far left)*, dating from the first century AD. Maps were then prepared for administrative purposes, using cartographic tools such as the folding rule and the proportional compasses *(left)*, also from the first century AD. The gap between the lower points of the compasses is twice that between the upper, enabling a map to be copied at half or double its scale.**

Engineering skills were also utilized in mining for gold, silver, copper, lead and other metals. In open mines, torrents of water delivered by aqueducts – as much as 34 million litres a day at one site in northern Spain – were directed to loosen and break down deposits, and series of sluice gates were used to process the waterborne debris. Trenches some 10 metres deep were also cut for surface mining. To gain access to subterranean sources of metal, Roman engineers cut deep shafts and lined them with wood or stone – these led to a network of underground galleries that followed the direction of the vein to be mined. "Not only do they go into the ground a great distance," wrote the chronicler Diodorus Siculus during the first century BC, "but they also push their diggings many stades in depth and run galleries off at every angle, turning this way and that, in this manner bringing up from the depths the ore which gives them the profit they are seeking."

Underground lighting was provided by oil lamps, which could be placed in niches cut in the walls or on ledges. Water that flooded the galleries was removed either by the manual labour of thousands of slaves or, in some of the more wealthy mines, by means of water wheels or massive Archimedean screws that drew up water as they rotated – "an exceptionally ingenious machine," declared Diodorus Siculus, "by which an enormous amount of water is thrown out, to one's astonishment, by means of a trifling amount of labour."

Order and discipline were brought to bear on the surface of the land with similar rigour. In northern Italy and certain other parts of the empire, most publicly owned

land was marked out by roads or paths intersecting at right angles into rectangles 2,400 Roman feet (700 metres) long; these plots were then divided into individual farmholdings of between 10 and 40 hectares. Villas – essentially farmsteads of above average size and sophistication, often with their own facilities for oil and wine production, weaving, dyeing and metalworking – replaced more basic agricultural settlements around major towns in the provinces. And after the North African city of Carthage was taken over in around 200 BC, the wide belt of fertile soil between the Mediterranean coast and the Sahara Desert was rapidly exploited to produce grain for export to Rome. Much land was divided up into a grid system of 50-hectare squares, which were settled and farmed by military veterans. Woodland was cleared, pastures and orchards ploughed over. "Cultivated fields have become the forests," observed the Carthaginian theologian Tertullian. "The sands are being planted, the rocks hewn, the swamp drained."

The timber from the forests that yielded to the plough was itself a highly prized commodity. Wood was virtually the only fuel available for heating, cooking and the smelting of metal ores; it was used extensively for constructing houses and making furniture; and, above all, it sustained military conquest. The Roman natural historian Pliny the Elder recorded that during the first war against Carthage a fleet of 120 wooden ships was built in 60 days, and for another naval campaign 220 ships were made in just 45 days. They were destroyed almost as quickly as they were turned out: according to another Roman historian, 700 quinqueremes – massive galleys with five banks of oars – were lost in the course of the conflict with Carthage. Armies were even more voracious in their appetite for timber: it was used for building bridges, roads, defence works and siege towers as well as for fuel. One tower erected to storm the walls of Rhodes was described in detail by a chronicler: nine storeys high, its top platform was 45 metres above the ground, and it required a force of more than a thousand men to move it into position.

Meanwhile, as the empire expanded, so too did the capital, whose population grew to more than a million by the first century AD. In 123 BC the civic authorities had taken on the responsibility of distributing a free ration of corn every month to all eligible citizens: slaves and those who did not qualify for the free handouts were supplied by private merchants. The poorer inhabitants of Rome had therefore come to expect guaranteed supplies of food, and satisfying this demand became a political problem of increasing urgency. In around AD 50 the Emperor Claudius was attacked in the forum by a hungry mob who pelted him with pieces of stale bread. Recognizing that here was a situation as volatile as any on the empire's remote frontiers, Rome's rulers determined to buy off discontent with cheap bread and cheap thrills. Both elements in this policy of bread and circuses intensified what had now become a coordinated assault on the natural resources of the provinces. To keep the bellies of Rome's citizens full, wheat had to be supplied fast and in bulk. And to keep their minds off rebellion, animals – the wilder and the more exotic the better – had to be captured for bloody gladiatorial spectacles.

Most of Rome's corn was supplied from Sicily, southwestern Spain and North Africa. Their distance from Rome was not a severe impediment, for despite the potential hazards of storms and piracy it was still both cheaper and quicker to transport grain by ship than overland. But the drain of the Romans' demands on the agricultural resources of the provinces proved so great that many of their own inhabitants were left without sufficient food. Rome was consuming more than 6,000 tonnes of

**This third-century mosaic from a Roman villa in what is now Algeria shows frantic leopards and lions being driven towards a net securely attached to stakes. Advancing upon them, the hunters thrust their flaming torches from between their shields; on the left, a reckless footman has been caught and felled by a leopard. The purpose of such hunts was to capture animals live for transportation to Rome, in whose thronged arenas they were pitted against gladiators in bloody, spectacular contests. The local extinction of the elephant, the rhino and the zebra in North Africa and the hippopotamus from the lower Nile in Egypt was largely due to the Roman craze for imported exotic beasts.**

Standing at the helm of his boat, a captain supervises the loading of a cargo of grain in this third-century mosaic from Ostia at the mouth of the River Tiber. In the first century AD some 300,000 tonnes of grain were carried each year to Ostia by seagoing freighters, chiefly from North Africa and Egypt; from the port's warehouses a shuttle service of lighters sailed 35 kilometres upstream to Rome. The coin adorned with six ears of wheat, dated around 20 BC, commemorates the reorganization by the Emperor Augustus of free supplies of grain for Roman citizens. His successor Tiberius warned the Senate that if this corn dole were discontinued, "the utter ruin of the state will follow".

corn per week at the beginning of the first century AD. To keep pace with this level of demand at least 800 shiploads of wheat had to reach the port of Ostia on the River Tiber during the short sailing season, which generally lasted from May to September. During the following century the Greek physician Galen complained that "the city dwellers, as it was their custom to collect and store enough corn for all the next year immediately after the harvest, carried off all the wheat, barley, beans and lentils and left what remained to the country people." The latter finished off their pulses during the winter months, "and so had to fall back on unhealthy foods during the spring; they ate twigs and shoots of trees and bushes and bulbs and roots of indigestible plants: they filled themselves with wild herbs and cooked fresh grass."

Animals were also imported to Rome from abroad, and in such large numbers that many of the larger species became extinct in those regions which were accessible to professional hunters. The wealthy maintained a ruling-class tradition of keeping certain animals as pets: the Emperor Hadrian went so far as to erect tombstones over the graves of his favourite dogs, and many aristocrats kept monkeys, birds and even fish as domestic playthings. But as far as ordinary Romans were concerned, animals were for simply providing meat and milk, hauling wagons, dragging ploughs – and, now, for entertainment. Before the huge crowds massed in the Circus Maximus and other stadiums, bears, elephants, lions, leopards, crocodiles and rhinoceroses were either slaughtered by professional fighters or goaded into destroying one another, having first been drugged with intoxicants which served to heighten their frenzy. Those beasts that survived the ordeal were then shot with arrows by spectators who had paid extra for the privilege of their ringside seats. The Emperor Nero had 400 bears and 300 lions killed with javelins by his personal bodyguard. To celebrate his military victories the Emperor Trajan had 11,000 beasts publicly butchered. Finer Roman minds were appalled by the spectacle: "Can't you people see even this much, that bad examples recoil on those who set them?" asked the philosopher Seneca. But the policy amply served its immediate purpose, and the populace of Rome, gorged and hoarse, went peacefully to bed.

Warfare and agriculture, the great twin engines of empire, fed ceaselessly upon each other. Since the large estates in Italy, Sicily and latterly North Africa were also worked by captive slaves – the great markets of Capua and Delos were capable of handling around 20,000 slaves per day – the dispossessed smallholders could not even sell their labour. They therefore drifted towards the urban centres, thus increasing the demands that each city placed upon its rural hinterland. Moreover, the ready availability of slaves discouraged the development of new agricultural techniques – so that output could only be augmented by further military conquest, leading to an extension of the area under cultivation.

But this apparently endless cycle in fact contained its own constraints. Firstly there were the simple limits of human ingenuity: the Romans proved incapable of adapting their high-yield agricultural methods to the wetter climate and heavier soils they found in central and western Europe. And secondly there were the limits of the land itself. Cleared, mined and constantly reploughed, the soil was showing increasing signs that it was losing its fertility. Sometime around 250 AD the Christian bishop of Carthage sent a grave warning to the Roman proconsul in Africa: "You must know that the world has grown old and does not remain in its former vigour. The rainfall and the sun's warmth are both diminishing; the metals are nearly exhausted; the husbandman is failing in his field . . . springs which once gushed forth liberally, now barely give a trickle of water."

For those with keen eyes the evidence of this decline had been there to see at least 300 years earlier. In Italy, the deforestation of the Apennines during the wars the Romans fought against Carthage had allowed soil to be washed down the hillsides,

**This third-century-BC Greek terracotta model of a camel bearing amphorae of wine was found in Syria, which exported grain to the populous city-states of Greece and received in return Greek wine, oil and pottery. The fastest and cheapest method of transporting produce was by ship, but along caravan routes across the hot, dry regions of Asia and the Middle East the camel came into its own. The animal's ability to store fat in its hump and to conserve body water, as well as its high tolerance of dehydration, enable it to travel for many days on a sparse diet of thorny bushes or dried grasses. These attributes made the camel an ideal pack animal for the desert plains of North Africa, where it was introduced around the third century AD.**

silting up rivers and turning coastal areas into marshland – including the Pontine Marshes at the mouth of the River Tiber. In around 60 BC the poet Lucretius observed that farmers were having to work harder than their ancestors had done to harvest the same amount of produce. But it was in North Africa and Asia Minor, the regions of the Mediterranean with the lowest rainfall, that the effects of overexploitation of the land were most conspicuous. Here, lush forests had once absorbed the precious winter rains and retained life-giving moisture in the soil around their roots. Now, with the trees gone, the soil exhausted, and supplies of dung too little or too distant to be brought to replenish its fertility, the wind simply whipped away the dry topsoil and the land began to turn to desert.

The Romans did take some positive steps in an attempt to conserve their resources. Marshland was drained in Italy, and terraces were carved in the hillsides of North Africa to hold rainfall or melting snow, which seeped slowly through the soil to fill the water table and replenish wells. In Rome during the first century AD there was a lively market for recycled glass: citizens collected fragments from the streets and sold them to local workshops, where less fuel was required to recycle the old product than was needed to make new glass from raw materials. At Ephesus, the major city of Asia Minor, where alluvium washed down from the hills by the River Cayster constantly threatened to silt up the port, the Romans repeatedly dredged the harbour and even made attempts to divert the course of the river. But all such measures provided only a temporary alleviation of the problems.

Moreover, there were ominous signs that the organizational genius and energy of Rome were beginning to wane. "No addition whatever is being made to knowledge by means of original research," complained Pliny the Elder, "and in fact even the discoveries of our predecessors are not being thoroughly studied." Many of the ordinary citizens of Rome, especially among the military and lower classes, were dismayed by the manner in which the official Roman religion had become increasingly deeply intertwined with the political protocol of the state – some emperors even decreed that they themselves were gods – and switched their loyalties instead to one of the variety of ecstatic mystery religions that were flooding into Rome from the east. Among these were the cult of Isis, a version of sacrificial bull rites that had flourished in Egypt centuries before, and a cult that derived from the ancient Indo-European deity Mithra, which also involved the ritual sacrifice of bulls. A third competitor for the allegiance of those Romans who decided to turn away from their traditional gods was Christianity, which despite the fierce persecution its followers suffered made great strides among the general populace. During the fourth century Christianity was adopted as the official religion of the Roman empire – a dangerous moment, for the new faith was openly hostile towards the class system and the rigid social hierarchy, built to a large extent on a foundation of slave labour, on which the traditional fortunes of Rome were based.

There were dangers from without, too. For centuries the Romans had been menaced by the nomadic raiders who lived on the open plains across the Danube and preyed on the northern frontier. *Barbari*, the Greeks had called them, lampooning their incomprehensible and barking language. During the fourth century these peoples began to advance south and west, swarming over the imperial defences and into France, Spain, Italy and North Africa.

Rome during its senescence, ruled over by a succession of weak emperors and succumbing increasingly to a mood of fatalistic pessimism, was no longer able to offer anything more than scant resistance. In 476 Rome's control over the western empire, which for some time had been barely more than nominal, was officially terminated.

Gradually the inhabitants of the entire Mediterranean region and western Europe came to realize that an era had ended. Merchant fleets packed tight with grain no longer converged on the wharves of Ostia at the mouth of the River Tiber. Mine shafts caved in. Water ran dry in aqueducts, and weeds grew up in the precincts of abandoned temples and on roads once trodden by the marching feet of legions. To many witnesses it must have appeared as if the land itself, along with the power of Rome to which it had been harnessed, had grown old and impotent.

Even more gradually, however, the land started to reassert its own rhythms of growth, as if to prove that all along it had retained its essential independence – and though borrowed by Rome and put to Roman uses, it was itself no more Roman than it had been Minoan or Mycenaean or Greek or Phoenician. Some regions would never return to their former fruitfulness: the more eroded parts of North Africa became home not to settled farmers but to nomadic pastoralists, whose herds of goats and other foraging animals prevented new tree cover from growing. But most parts of the Mediterranean, no longer subject to the overweening demands of Rome and its provincial capitals, now had to supply more immediate wants, and proved equal to that challenge. Rivers still flowed in a seasonal cycle of spate and decline; olives and vines and wheat and barley and a variety of vegetables nourished the people of villages and small towns.

Lapsed from a severe and uniform regimen, the resilient Mediterranean landscape became a mosaic of varied and productive local agricultures – as it remained until the 20th century's technology and mass demands imposed their own exacting monocultures. Even these did not trespass – for by then there was profit in tourism, and a judicious sense of limits – on the hollow stone ruins of Greece and Rome.

**A page from an 11th-century Arabic edition of a Greek catalogue of some 600 plants and their medicinal uses shows the agrimony plant, which was used to cure liver ailments. The inventory was compiled in the first century AD by Pedanius Dioscorides, one of a number of Greek scholars who rejected magical and divine explanations of diseases in favour of empirical observation of symptoms and cures. Later translated into at least seven other languages, Dioscorides' work served as the primary text of pharmacology in Europe until the end of the 15th century.**

# THE CHINESE WAY

3

During the reign of the legendary Chinese emperor Yao, 10 suns shone in the sky at the same time, causing rivers to dry up and the earth to become barren. By dint of blind perseverence and determination, the hapless people of this drought-stricken land managed to eke out a meagre livelihood – but just as they were beginning to prosper, a great flood swept over the world and reduced them again to misery.

So begins, unpromisingly, an account dating from several hundred years BC of the beginnings of Chinese civilization. But the story goes on to tell how, with divine help, the destructive forces of nature were eventually tamed, and order and control were brought to the land. A god named Gun took pity on the suffering people and, using a piece of soil with magical properties, built a dam to hold back the flood waters. Gun was killed by the angry God of Fire, and again the waters covered the land. But Gun's spirit refused to die, and after three years a son, Yu, emerged from his corpse. Yu was able to obtain the services of a winged dragon and some more of the magical soil, which he used to stop up the 233,599 springs from which the water was emerging.

Yu next turned his attention to the floods that still covered the landscape. For 13 years he travelled by boat on water, by sledge over marsh and by cart on land, and did not stop to enter his own house even though he passed by the gate. In this way he made a survey of the land and marked out the boundaries of nine provinces; he also cut deep channels that became the great rivers of China, and made passes through the mountains and built dykes in the low-lying plains. Yu also fathered a son, Qi, who became the founder of the first hereditary dynasty in China.

Floods and drought have repeatedly punctuated Chinese history, and Yu's efforts to bring stability to this troubled land speak for the experience of countless generations of ordinary Chinese people. Their successes have always been precarious: famine has so often cancelled out the labour of hundreds of thousands of families that for most Chinese, for most of the time, bare subsistence has been the only realistic goal. But their achievement, however hard won, is undoubted. The Chinese now have a life expectancy – 68 years at birth for males and 71 years for females in 1988 – that is higher than in the U.S.S.R., India and any nation in Africa, and comparable with the richer nations of the Western world, themselves less subject to natural calamities.

In coping with the natural disasters that frequently beset their land and with the economic problems of feeding an enormous population, the Chinese and other peoples of eastern and southern Asia have resorted to methods of agriculture very different from those characteristic of the West. From at least the beginning of the 19th century, America and Europe have fed their own citizens by the ever more intense application of industrial technology. In China and India, various other strategies – determined by the land itself and by cultural attitudes towards it – have prevailed. The effects have been most apparent in the biological sphere: the introduction of new

**A detail from a 16th-century Japanese screen shows a team of women planting rice in a flooded field, where the seedlings will remain until the field is drained in preparation for harvesting. Wet cultivation of rice was established in China possibly by the end of the fifth millennium BC, and spread to Japan around 500 BC, and also to India and many parts of Southeast Asia. Rice growing, although one of the most highly labour-intensive forms of agriculture, is also one of the most productive: rice is the chief food for three out of five of the world's inhabitants. In many Eastern countries handfuls of rice are often thrown over newly married couples to wish them fertility, a custom from which the Western practice of scattering confetti derives.**

varieties of crops, many of which have been specifically developed for growing on inferior land, and of labour-intensive systems of agriculture that make efficient and economical use of natural resources. Both China and the Orient have prospered not so much by mastering nature as by the patient expenditure of infinite amounts of human energy and a minute attention to detail.

Present-day China is the third largest nation in the world and is home to more than a billion people, or a quarter of the earth's population. But such enormous statistics conceal an immense variety of terrain, of peoples and of cultural traditions. If China

**The hand of the farmer has transformed this landscape near Yangshao in southern China into a patchwork of fields interlaced by irrigation canals. Because of the pressure of supporting a vast population, every cultivable piece of land up to the edge of the sheer-rising mountains is used for producing crops: rice in the flooded paddy fields, and wheat, beans, vegetables and tobacco in the dry fields.**

exists at all as a unified entity, in the minds of both outsiders worldwide and of the Chinese themselves, it is only because of a lengthy and gradual historical process that involved the development of complex and resilient forms of political control, an eclectic but unifying culture, and methods of working the land that rely above all on cooperation and shared endeavour.

Astronauts orbiting the earth have reported that the only man-made construction visible from that distance is the Great Wall of China, the barrier that was erected along the northern frontier to protect the settled Chinese from their nomadic, aggressive neighbours. As they tumbled towards earth and the surface of the globe came into ever sharper focus, the astronauts might also have noticed that China comprises two quite distinct geographical regions. Inner China, which forms a roughly square-shaped area bordered to the south and east by sea, is a land of alluvial plains and river valleys, a patchwork of tiny fragments of every shade of green and brown. The historical core of the Chinese empire until the 17th century, this terrain is home to about 95 per cent of the population. On its north and west sides it is wrapped around by outer China, a region of greatly contrasting geographies. Outer China includes the world's highest mountain, Mount Everest, towering a mighty 8,848 metres above sea level, and the second lowest place on land, the Turfan Depression, which lies 154 metres below sea level. It encompasses, too, the burning, arid deserts of the Gobi and Taklamakan, the rolling, grass-covered steppes of Mongolia as well as vast tracts of virgin forest and swampland. Not surprisingly, large areas are uninhabited, and the overall population density does not exceed one person per square kilometre. Recent centuries have seen the migration of substantial numbers of people from inner into parts of outer China, notably Manchuria, but these peripheral regions are still primarily the homelands of the Mongols, Tibetans and other peoples who make up 54 of the 55 ethnic nationalities that are recognized by the People's Republic of China.

Simple human habitations nestle by the shore of a lake in this scene from a scroll painted in ink by the 17th-century Chinese artist Wang Hui. Rejecting realism, traditional Chinese painting favoured harmonious, idealized compositions in which the pulse of nature was expressed in the forms of surging waterfalls, lofty peaks and branching trees. Such works embodied reverential attitudes to the natural world that coexisted with its necessary and increasing exploitation.

Inner China in turn divides into two very distinct areas. From the mountainous region lying to the west, the Qinling and Daba ranges of the Kunlun mountains extend like a long finger and thumb to grasp the trailing line of the Han, a tributary of the Yangtze River. This river, and the mountains it threads into, define the boundary between quite different climatic and geographical zones.

South China – as also the whole of Southeast Asia – ranges from the temperate to the subtropical in climate. In the Sichuan basin and towards the South China Sea, both temperatures and rainfall are high, and the growing season lasts nearly the whole year long. In ancient times the land was all but covered by dense forests of pine and bamboo, which were home to creatures such as elephants and giant lizards; the human inhabitants occupied the coastal lowlands and valleys, and supplemented the produce of their simple agricultural way of life with hunting and fishing. Farther north in the Yangtze Valley, rainfall is lower, and there is a tendency

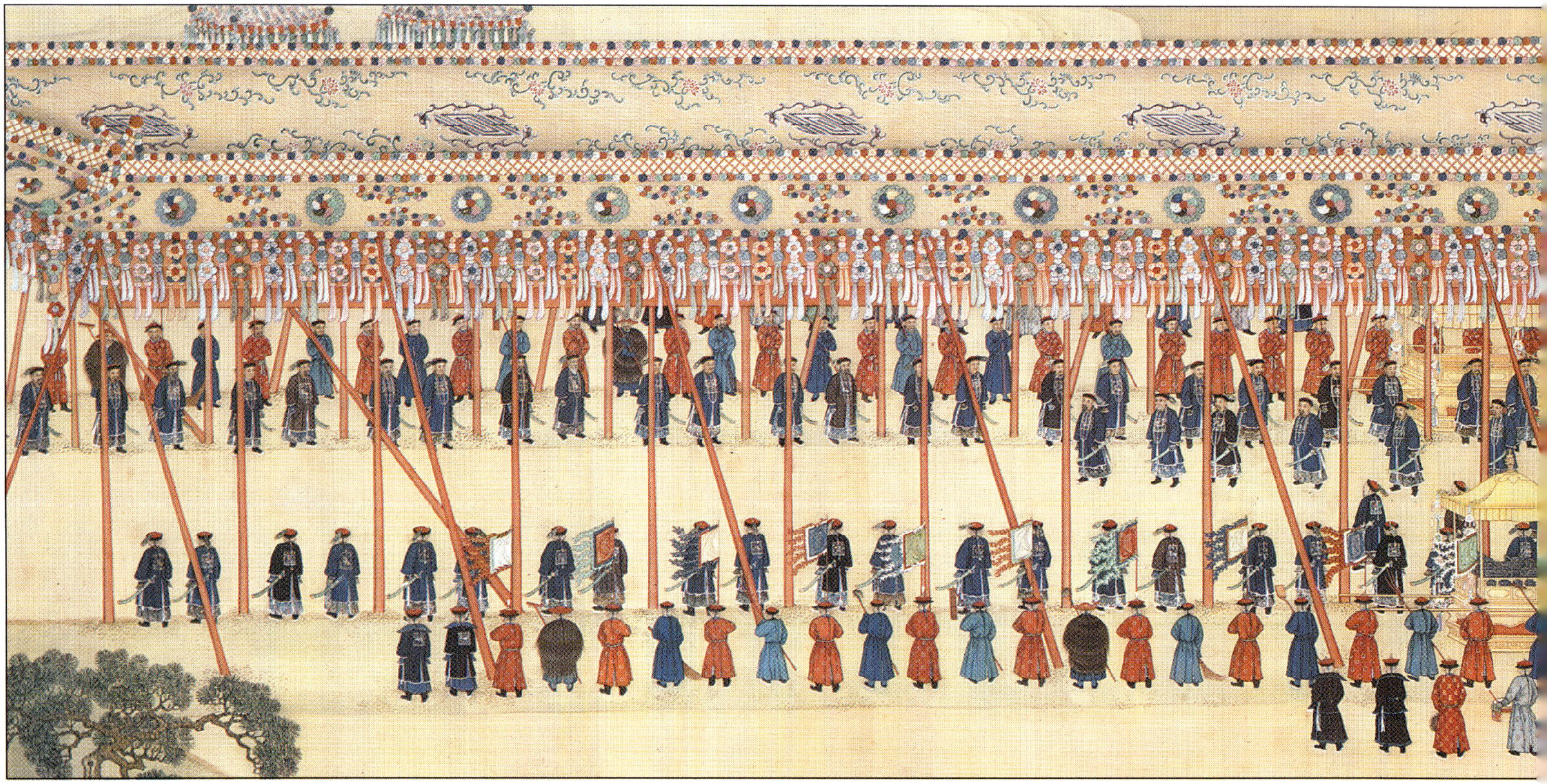

An 18th-century scroll painting on silk shows ranks of courtiers beneath an elaborate canopy attending the Emperor — descended from the dais on the far right — who is about to plough the first furrow of the year. Parallel with the Emperor and awaiting his lead, high ministers stand behind their own ploughs. In addition to performing this ceremony — intended to ensure a successful growing season — Chinese emperors since the establishment of the earliest dynasty were expected to sacrifice each spring and autumn at the altar of the spirit of the soil. These customs suited imperial interests: good harvests lessened the likelihood of rebellion, and most of the emperors' revenue derived from taxes on land and produce.

towards dry winters. The summers are hot and long, and the growing season lasts as long as nine to 10 months. Ban Gu, an historian from the first century AD, wrote an observant, if subjective, description of the sort of life style enjoyed by the denizens of this region during the fourth century BC: "The people live on fish and rice. Hunting, fishing and wood gathering are their principal activities. Because there is always enough to eat, they are a lazy and improvident folk, laying up no stores for the future, so confident are they that the supply of food and drink will always be replenished. They have no fear of cold and hunger; on the other hand, there are no rich households among them. They believe in the power of shamans and spirits and are much addicted to lewd religious rites."

To the north of the Yangtze lie wide plains through which the Yellow River, the second of China's two great arteries, passes. The plains are covered by a fine, yellow loess soil that was picked up by strong winds from the deserts of central Asia during the glacial period and deposited over the northwest of China, sometimes to a depth of several hundred metres. Loess soil is easily worked and fertile, and about 500,000 years ago would have been pierced by outcrops of forest-covered rock, a source of both timber and water. Here, where the Yellow River emerges from the mountains but is still protected by hills, the first Chinese people lived.

By around 5000 BC settled communities of farmers were living in villages in the regions of present-day Shanxi and Henan provinces in northern China and Longshan, farther to the east in Shandong. Their houses had walls of wattle and mud and floors of beaten earth. They had domesticated dogs and pigs, and had developed sophisticated techniques of making and decorating pottery. Using implements of stone and wood, they cultivated millet, wheat, beans, rice, hemp, cabbages and

melons. Gradually, agricultural prosperity led to the development of cities, the first of which were most probably established by clans headed by warrior-leaders whose dominance was ensured by their possession of bronze weapons and horse-drawn chariots. The Shang kings of the 17th to the 12th centuries BC indulged in great hunting expeditions as preparation and training for their endless forays into battle against other cities and against the less-developed peoples who occupied the surrounding countryside – the remains of elephants, rhinoceroses, bears, tigers, leopards, deer, monkeys, foxes and badgers have all been unearthed on their hunting grounds. The staple crops of wheat and millet supported a growing population, and the kings were able to call upon the services of scribes and other officials who were not directly involved in the production of food.

As the use of bronze technology expanded, peripheral regions were settled and brought under cultivation, and the ever-increasing grain production supported larger armies. One by one any smaller and weaker states were subsumed until, in 221 BC, the state of Qin commanded a unified Chinese empire for the first time. The Qin state sought to establish its authority with systematic vigour: texts deemed to be subversive were proscribed, weights and measures were standardized, travel outside the state was allowed only with a permit, and armies of corvée labour were drafted to undertake a massive programme of irrigation and road building works. But the death of the first emperor was soon followed by rebellion, and in 206 BC a new dynasty, the Han, was established. The next 400 years of the Han dynasty saw the expansion of Chinese culture to a large part of eastern Asia.

Large numbers of people had already migrated south, fleeing wars or natural disasters that threatened the stability of life in the Yellow River homelands. Around the fourth century BC, over one third of the registered population of North China, approximately two million people, had moved south of the Yangtze; and in the wake of mass migration to well-endowed regions such as the Sichuan basin, people were further displaced to the south and southwest. The Han rulers encouraged such movements of population: seeking to maintain control of the trade routes to the west and to central Asia, they established self-sufficient colonies of farmer-soldiers in the fertile river valleys, and settled as many as 700,000 colonists in the Gansu corridor in the northwest at the end of the second century BC. Those peoples who did not submit to Chinese rule were pushed out of the cultivable lands and forced to adopt a pastoral, nomadic way of life on the steppes beyond.

Trade filled the Han coffers to overflowing: the Roman empire's insatiable demand for supplies of silk and Chinese spices ensured a continuous stream of merchants along the caravan route through central Asia known as the Silk Road, the balance sheet being very much in favour of the East. Agriculture too brought a new measure of stability. Zhao Guo, an official employed at the Han court during the first century BC, recommended ploughing wide and deep furrows; when the seedlings sprouted, the soil from the ridges was then piled up round them until the level of the field was even. Such methods helped to make harvests more secure, and were subsequently put into practice throughout the empire.

At the same time, a variety of intellectual and religious influences were shaping man's distinctive cultural attitudes towards the natural world. Among them were the ancient concepts of Yin and Yang – two complementary but opposing forces that represent darkness and light, female and male, weakness and strength. Yin and Yang were believed to rise and fall cyclically, a theory that was used to explain the pattern of birth, death and rebirth in the natural world, and the progression of the seasons. All human and animal activity was believed to have its optimum point in the cycle, and to carry out an activity outside the appropriate time spelt certain disaster. However, the influence of Yin and Yang could be channelled by judicious human intervention. In his account of his travels in China during the ninth century the Japanese monk Ennin reported the following traditional method of controlling the climate: "When, seeking good weather, you block the north road, Yin is obstructed and Yang then pervades and the skies should clear. When asking for rain, you block the south, Yang is obstructed and Yin then pervades and rain should fall."

The teachings of the philosopher Confucius, who lived in the sixth century BC, provided justification for the exercise of authority both in the state and in the family by concentrating on form and ritual. Ideal standards of conduct, for everyone from rulers of state to children, were set down, and exemplary individuals were held up as models. From a Confucian point of view, natural disasters and political downfall were both seen as the result of a failure to observe the correct rites.

A very different way of understanding the world counterbalanced the somewhat conformist doctrines of Confucius and his followers. Laozi, who was believed to have lived during the same era as Confucius, and his successor Zhuangzi, were interested in attaining freedom from man-made restrictions and rules, hoping thereby to achieve enlightenment – that is, an understanding of the Dao, or Way, the principle underlying the universe and all creation. Those in search of the Dao were often inspired to become hermits, finding a solitary life in a mountain fastness more conducive to contemplation on such matters than the hurly-burly of society. Such a life was portrayed by generations of Chinese painters in their pictures of mist-shrouded mountains on which the tiny figure of a white-haired scholar gazes from the shelter of his rustic pavilion. The fifth-century writer Kung Zhigui described the ideal towards which Daoist recluses should aspire: "A man who, incorruptible, holds himself aloof from the vulgar; untrammelled, avoids earthly concerns; vies in purity with the white snow, ascends straightaway to the blue clouds."

Later Daoists claimed to have received revelations of sacred texts, and Daoist cults that involved religious healing achieved a popular following. Seeking to escape the penalties of mortality, some Daoists put their interest in natural phenomena to service in the endless quest to find the elixir of life – although for those whose chemical experiments made use of poisonous materials such as cinnabar, this search sometimes brought about the exact opposite of the desired result. Daoist priests, who were based in temples situated in remote parts of the countryside, collected and categorized all manner of herbs and medicinal plants, compiling important pharmaceutical natural histories. Often they noted some surprising – and unproven – associations between various natural phenomena: "When in the mountains there is the *xia* plant [a kind of shallot]," according to a Daoist treatise that was written about AD 800, "then below gold will be found. When in the mountains there is the ginger plant, then below copper and tin will be found. If the mountain has precious jade, the branches of the trees all around will be drooping."

Daoist priests became practised in the art of geomancy – known in Chinese as *fengshui*, "wind and water" – which ensured that tombs and buildings were not sited in locations that disrupted the powerful forces believed to run through the landscape. Despite official disapproval, geomancers still operate in rural China.

A third body of thought that was to make a lasting contribution to Chinese attitudes to their world was of foreign origin. Buddhism was initially introduced to China

## PUTTING WATER TO WORK

**Both blessing and curse, China's waterways nourish the land and – by flooding adjacent plains at irregular intervals – threaten mass devastation. Since at least the third century BC armies of labourers have toiled to build irrigation canals and flood-proof dykes; their largest project was the construction in the seventh century of the Grand Canal, shown overleaf, whose primary function was to transport grain over a distance of about 1,000 kilometres from the south to the north. A detail *(right)* from a 17th-century scroll shows workers bringing a tied bundle of reeds to repair a dyke while others hurry by with mattocks and baskets of earth. Some overseers offered prizes of clothes, meat or wine to the fastest workers.**

during the first century AD by Indian missionaries who were following the traders on the Silk Road. It was first adopted among the upper echelons of Chinese society, and underwent substantial transformation as the belief became more popular and widespread. In China, for example, the Buddhist concept of the transmigration of souls was developed into ideas of paradise and the underworld that were quite alien to the Buddha's original teachings. But the promise of peace and plenty after death had a considerable appeal to a population whose earthly lives were both exhausting and perilous. An ancient adage neatly summed up the pragmatic attitudes of the Chinese towards their eclectic cultural traditions: "Confucian in office, Daoist in retirement, Buddhist as death draws near."

The first Chinese census was conducted in the year AD 2, and numbered the population of the Han empire at around 60 million. The borders were still insecure and the people within the empire followed many different ways of life, but the very fact that the census could be taken indicated that a strong central authority had been established over a very large area. A major contribution to this unity was the degree of communal effort required to bring the land under human control and make it bear sufficient produce for a growing population.

The first imperative was to harness the potentially destructive torrents of the major rivers. Vast armies of labourers were accordingly mobilized to erect dykes to confine and control the waters of flooding rivers, and to dig ditches to divert this resource to irrigate the soil. All the major rivers of China, but the Yellow River in particular, pick up large amounts of soil as they cut through the hills and mountains in their upper

reaches. This load is carried downstream until, as the flow of the river slows down on reaching the plains, it is deposited on the riverbed, thus raising the water level and causing the river to burst its banks. The Yellow River, known as "China's Sorrow", carries some 1,450 million tonnes of silt annually, and has changed its course dramatically over the past 2,000 years – from following a path almost due north, to debouching south of the Shandong peninsula, and then gradually moving back again to emerge where it does today in the Bo Hai gulf. In some parts of the alluvial plain, centuries of dyke building have resulted in the course of the river being raised several metres above the surrounding landscape.

A complex system of irrigation canals in the low-lying delta both evenly distributed water over the land and, by means of reservoirs, enabled a certain amount to be stored for times when rainfall was scanty. The building of irrigation channels and dykes had a further effect on the soil of the delta region: the silt that was dredged from the canals during periodic clearing was deposited on the fields, creating a layer of topsoil that could be as much as a metre thick.

The most ambitious project designed to put water to productive use was the construction of the Grand Canal at the beginning of the seventh century AD. The canal's primary purpose was to enable rice grown in the Yangtze delta region to be transported quickly and cheaply to the capital in the north, and to the soldiers stationed on the northern frontiers; because of the lack of a good road network, especially in the south, it was also used for many other forms of traffic. The canal began in the fertile fields of the Yangtze delta and then crossed the low-lying Huai River valley, incorporating conveniently placed tributaries in its route; a northeastern arm reaching as far as present-day Beijing served the garrisons there. It was built by a

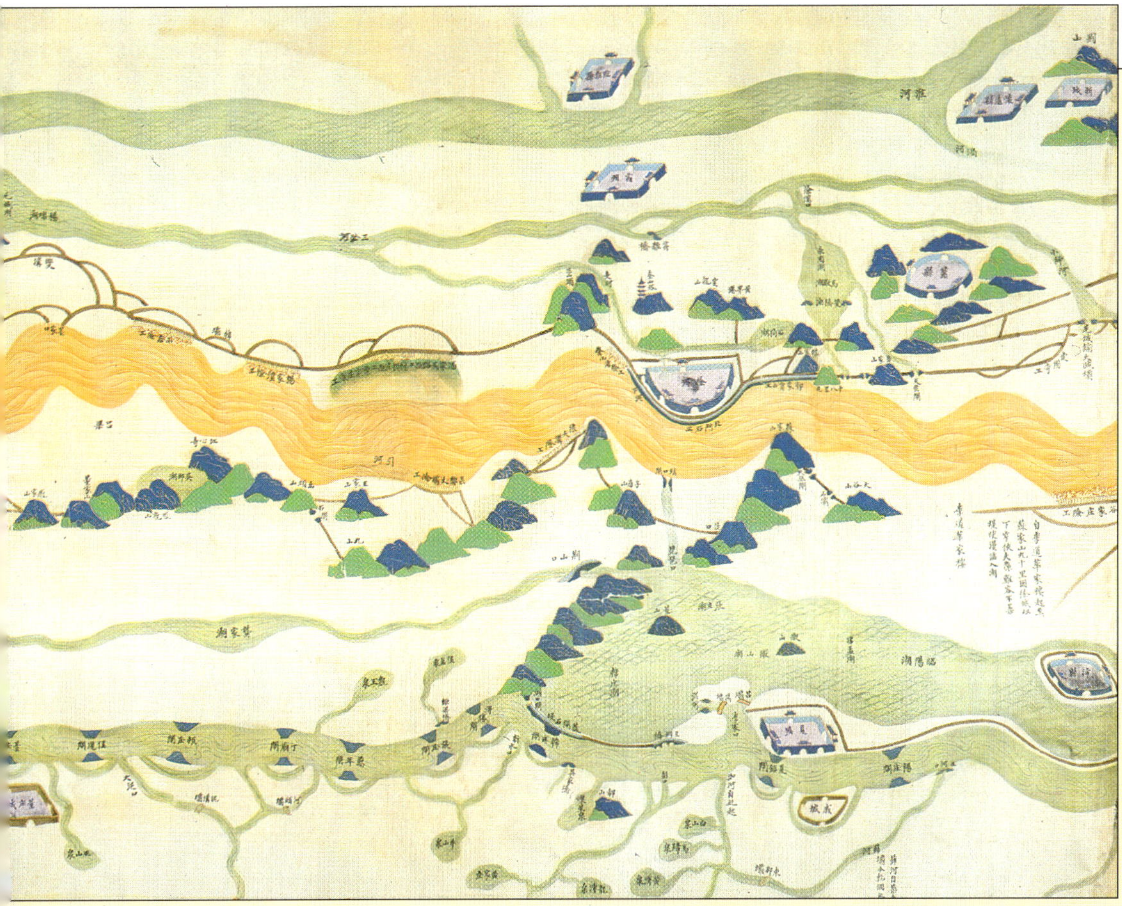

1 2 3

**A section from a 14-metre-long scroll map of the Grand Canal, drawn in the 18th century, depicts – on the left – the junction of the canal with Lake Luoma and the Yellow River, and the beginning of its southward course to the Yangtze River. That the canal is shown here to run parallel to the Yellow River *(coloured yellow)* is the result of a distortion of directional values that allows the map to include a mass of information of benefit to contemporary administrators. The boxed details above show symbols representing (1) dykes to prevent flooding, (2) the placement of gabions, the stone-filled bamboo cages used for repairing dykes, and (3) lock gates in the canal. Also recorded are mountains, springs, towns and bridges. The passages of writing are lists of villages and families responsible for the maintenance of specific dykes.**

conscript force of 5.5 million labourers supervised by 50,000 guards, and in the ensuing centuries hundreds of thousands of workers were drafted for periodic repairs.

Raising water from canals to irrigate the fields required the use of pumps, powered by the wind, animals or, more usually, people. Simple devices such as the well sweep, a counterbalanced bucket, and treadmills, which consist of a chain of paddles or buckets that carry or push before them a supply of water, are still in use today in many parts of China. A more sophisticated device introduced in the 10th century was the noria, a form of giant water wheel whose highest scoops spill water into a pipe that is carried on props over the adjacent fields. Where a fast-flowing river is the source, the noria can be self-propelling. A 12th-century poet considered these devices worthy of praise in verse:

*Ten tubes upon each wheel both drain and irrigate,*
*Rising and falling in a circle without cease.*
*Rich lands have been opened up in the surrounding hills,*
*Along ten thousand acres wind rice-sprouts like green clouds,*
*Clouds that before your eyes turn from green to gold,*
*And fill all stomachs with the year's rich harvest.*

The crop celebrated by the poet was grown mainly in the south, where the rainfall conditions were ideal, and as this region became more populated rice gradually supplanted wheat and millet, the more ancient staples of northern China. Indeed, it was to become the most plentiful crop of the entire world: one in 10 of the earth's arable hectares is now used for cultivating rice, which produces more than 650 Calories per day for every man, woman and child.

The highly efficient system of air passages leading from the leaves to the root of many varieties of rice plant enable them to be grown in waterlogged conditions that would drown most other crops. Some strains can survive under floods that are up to five metres deep, growing as much as 25 centimetres a day to keep their leaves above the rising waters. And far from being a handicap that the plant has adapted to overcome, water in fact promotes the growth of the rice plant: controlled flooding both prevents nutrients from escaping into the atmosphere or the subsoil, and makes for a highly efficient supply of nitrogen and phosphorus, the two most important plant minerals. Without artificial fertilizers, well-tended rice fields can produce two or three harvests each year for centuries from the same area of land.

Rice growing was and remains a highly labour-intensive activity. No mechanical device has the ability to sow the seeds in prepared beds, transplant them to paddy fields after about a month's growth, and then perform the weeding and harvesting, with the same efficiency as human hands. Each hectare of land under traditional methods of cultivation requires between 1,000 and 2,000 man-hours of labour to produce one harvest. Also, a high degree of human organization is needed to control the use of water in the paddies and on the hillside terraces: because the land must be dry when the rice is harvested and for a short time afterwards, growing cycles must be staggered, with parts of the land underwater while others are kept dry, to ensure continuous productivity. But the amount of labour and coordination required have themselves contributed to the stability and lasting success of this form of agriculture.

Where conditions determined, further sophistications were built into the system. If the growing season was not long enough to allow successive harvests, intercropping

**On a bamboo lattice, silkworms that have been fed on mulberry leaves spin the fine silk fibre secreted by their glands into shell-like cocoons. The candour of this painting, produced for export in the 19th century in a style designed to appeal to European tastes, contrasts with the jealousy with which China guarded its monopoly on silk production for more than 2,000 years. By the 11th century, sericulture – the art of raising silkworms – had spread to Japan and, via Arabia and India, to Europe.**

was practised; this involved a second planting of seedlings, which were set out in alternate rows a few weeks after the first. And in those areas where land suitable for cultivation was in short supply, plants that could be harvested without mutual interference were interwoven in rows. A Han dynasty text recommended that scallions should be grown under melon vines, and other works suggested combining onions and coriander, and turnips and hemp.

In order that the yields generated by such intensive farming techniques stayed high year after year, the Chinese peasantry developed elaborate methods of maintaining the fertility of the soil. Organic material that was dug into the soil to improve its nitrogen content included leguminous plants such as green beans, mud taken from rivers and canals during dredging operations, animal dung and night soil – that is, human excrement. Farmers augmented their own family's production of this vital resource by building latrines at the side of the road solely for the use of passers-by, to mutual benefit. In cities, night soil was collected each morning and transported out to the countryside where it was applied directly to the fields or else dried for later use. The sight of the euphemistically named "honey wagons" being pulled along city streets by horses or people is still a relatively common one. The cumulative effect of the nurturing of the soil over generations is evident in the high humus and nitrogen levels of the zones around cities.

In addition to the staple cereals of wheat and rice, the peasantry grew a range of other economic crops. A guide for the improvement of rural life written in the sixth century AD gave advice on the cultivation of vegetables, fruit, fibre crops and trees for timber; it also included recipes for food, medicines and wines, and instructions on how to make such daily necessities as soap, glue and dyes. The treatise stressed the

The tensile grace of bamboo stems – celebrated in the poem inscribed on this 13th-century ink drawing – is just one of the plant's virtues. Ubiquitous in China, and widely grown in Southeast Asia and on the islands of the Indian and Pacific oceans, bamboo grows faster than any other plant – up to a metre per day. Young shoots are cooked and eaten as vegetables; the pulp and fibre is processed to make paper; some species contain chemical compounds used in medicine. Bamboo is also used to make a vast range of domestic artefacts as well as in house construction and shipbuilding. Shown above, for example, is a section of a 230-metre-long bridge over the Min River in Sichuan that is made entirely of bamboo.

importance of being guided by the weather and the type of soil available: "Following the appropriateness of the season, consider well the nature and conditions of the soil, then and only then least labour will bring success. Rely on one's own idea and not on the orders of nature, then every effort will be futile."

Subsidiary crops have included maize, kaoliang – a type of sorghum – sweet potatoes, and peanuts. Tea bushes were grown on hillsides throughout South China, where variations in soil and climate produced distinctive flavours in different districts. Soya beans proved a rich source of protein and were also important in crop rotation to improve the soil. Hemp and ramie, a type of nettle, were the plants from which the clothes worn by most Chinese were made until the 14th century, when cotton was introduced from India and Southeast Asia.

For centuries China's most profitable export was silk, which also – as a woven fabric or as wadding for padded garments – clothed the Chinese upper classes. Silk is derived from the larvae of the silkworm moth, whose staple diet consists of the leaves of mulberry trees; the worms were domesticated at least 3,000 years ago, and China enjoyed a monopoly on the production of silk until the middle of the sixth century AD, when two Persian monks returning from China succeeded in smuggling some moth eggs and mulberry seeds into Byzantium. Although modern instruments such as humidity gauges have made it easier to control the environment in which silkworms are raised, the process has remained largely unchanged. The silkworm moth eggs are to begin with exposed to harsh weather or dipped in brine to kill off the weaker specimens. On hatching, the larvae are fed constantly for about a month on just mulberry leaves; during this time they surround themselves with cocoons of a delicate fibre, itself produced by liquid secretions that harden on contact with the air. When the larvae finish spinning their cocoons, they are plunged into boiling water, which instantly kills the new moths and loosens the fibre. As much as 900 metres of thread can be reeled from each cocoon.

Because of the constant pressure on land for cultivation, only marginal terrain was devoted to grazing and very little of the grain was expended on animal feed. Animals that could be fed on household scraps, or that could scavenge for themselves, such as pigs and poultry, were the livestock most commonly raised for food. The strong, hardy, water buffaloes proved ideally suited for pulling ploughs in the warm, damp conditions of the south, but they were a luxury that was beyond the means of most farmers. Limited numbers of horses and mules were used for transportation; usually it was cheaper for humans to haul or carry goods.

A similar sort of situation prevailed in India, where a vegetarian diet was the norm for the great majority of the population. Indeed, for the adherents of Buddhism and Jainism in India, vegetarianism was not just a necessity of life but an expression of virtue: according to the doctrine of the transmigration of souls, even the smallest insect might harbour what had been and could again become a human soul, and which should therefore not be killed. In India, however, more use was made of dairy products, especially ghee – boiled and clarified butter – which rapidly became an essential cooking medium. Vegetables or lentils were cooked in ghee, flavoured with spices and then diluted with coconut milk to make a curry sauce that turned a simple bowl of rice into a substantial meal.

The scarcity of meat did not restrict the potential variety of the Chinese diet. A poem written in the second century BC celebrated a banquet that included pigeon, goose, chicken, oriole, roast crane, magpie, bream, turtle, pickled pork, dog cooked in bitter herbs, five varieties of grain and four kinds of wine. This was hardly everyday fare, but for centuries the filling staples eaten by the ordinary Chinese have been enlivened by a combination of less copious dishes based on vegetables, eggs or meat. The greater the profusion of the latter, the more luxurious the meal.

Traditions of local cuisine were established which closely reflected the geography and consequently the produce of each region. In the grasslands of Mongolia wheat bread is staple and meat, particularly lamb and mutton, is plentiful; dairy produce is also very popular. In the time of the Tang dynasty, which lasted from the seventh to the 10th century, the taste for milk among the population was widespread, but most Chinese today find it repulsive. The northern Chinese consume vast quantities of steamed bread, wheat noodles and stuffed dumplings. Sichuanese cooking is characterized by the liberal use of chillies, a vegetable which was introduced from Central America in the 16th century; in the Yangtze delta fish forms an important part of the diet, and sweet flavours are popular. Ingredients are often cut into shreds the size of matchsticks, and are usually cooked quickly at high temperatures – probably because of the shortage of firewood for fuel.

Far from being the parochial customs of isolated provinces, local traditions of agriculture and cuisine had by the 14th century become distinctive strands woven into a huge and productive society. Following the introduction of the fast-maturing Champa strain of rice from Vietnam in the 11th century, which ripened in a third less time than did the native plant, the land could be made to yield two, or even three harvests each year. Yields were also improved, it was discovered, by the careful preparation and conservation of the soil. Transport benefited from the new paved roads and the use of lock gates in a spreading network of canals, and a nationwide system of customs houses levied tariffs on goods as they were moved from place to place. The growth of markets stimulated trade both within and among China's three main economic regions – the north, the lower Yangtze Valley and the Sichuan basin – and opened up a number of specialized occupations to the rural peasantry, including forestry, paper making and weaving.

In other fields also, China's technological achievements during the period known in Europe as the Middle Ages were vastly superior to those of the West. Machines for reeling silk and spinning hemp speeded the manufacture of clothing. Doctors learnt to diagnose diseases with a new accuracy. Coal was used in the smelting of iron ore; gunpowder revolutionized warfare; and the fields of mathematics and astronomy flourished. Visiting China towards the end of the 13th century, the Venetian traveller Marco Polo was filled with awe and wonderment at the bustling, teeming cities and the productive countryside. "Indeed it is scarcely possible to set down in writing the magnificence of this province," he declared; and of the city of Hangzhou he wrote that "anyone seeing such a multitude would believe it a stark impossibility that food could be found to fill so many mouths."

For all governments of East and South Asia, the problems of balancing the food supply with a growing population have been a pressing concern – not least because a hungry populace can quickly become an army of rebellion. The shelves of the Han imperial library were filled with scrolls on agriculture and medicine as well as on astronomy, mathematics and military matters. Very soon after the invention of printing in China in the ninth century AD, treatises containing advice on husbandry began to circulate widely. But only a tiny proportion of the population was able to read, and the spread of information could have only a limited effect. The most direct means of controlling production, as rulers soon realized, was through control of the land itself and the people who worked it.

Many forms of control were attempted. In northern China in the fourth century, for example, land was allotted to individuals for use in their lifetime, with provision made for officials to have larger portions according to their rank. But this system proved incompatible with the wet-field farming of the Yangtze Valley, as it failed to ensure that those who made the great investment of collaborative labour needed for irrigated farming to succeed were the ones who subsequently reaped the benefits. Other systems of land tenure varied from one region to another. In some places, such as the mountainous regions of today's province of Zhejiang, peasants were usually independent; in others they were virtually slaves, obliged to work under the close supervision of an overseer; in those areas ravaged by warfare or underpopulated for other reasons, the government often took direct control. In general, the more productive the land the more likely it was to be owned by landlords and worked by tenants.

Landlords clearly exercised immense control over the lives and fortunes of others, and the Song dynasty that ruled from 960 to 1279 developed a system of coopting members of this class in the administration of state. The wealthiest people in a village were required in turn to perform such duties as the collection of taxes, supervision of waterworks, arbitration of disputes, road maintenance and, where necessary, the distribution of famine relief. But even this system did not solve the perennial problems of government — for whenever the state devolved responsibilities to local elites, they would build on their position of power to challenge imperial authority. Many took

In Papua New Guinea the heads of butchered pigs are impaled above stones being heated to cook their flesh. For certain peoples in New Guinea and the South Pacific islands, the pig is a holy animal that must be sacrificed to their ancestors, and once or twice in a generation these peoples slaughter and eat the majority of their herds in a ritual feast. The roots of the practice may lie in the need to regulate the numbers of fast-breeding livestock in order that they do not encroach on the growing crops.

# THE LAWS OF DIET

**In some societies, animals deemed sacred are eaten; in others, their flesh is forbidden for the same reason. As the examples shown here and overleaf confirm, man's attitude towards food obeys no universal pattern, often because it has evolved in response to very specific local conditions.**

**When the early farmers discovered that a particular sort of meat caused illness, they avoided it. When the raising of a certain animal conflicted with their essential way of life — pigs, for example, which yield no milk, are less vital to pastoralists than sheep or goats — they learnt to do without its meat, however delicious. Over time, customary habits often developed into religious prescriptions, themselves reinforced by other factors — the need to standardize behaviour so as to maintain the social order, or to give expression to a people's sense of their unique identity.**

advantage of their position to evade their taxes, passing the burden on, through their managers and agents, to those in their sway.

From the 15th century onwards the administration of China was largely carried out by a class of educated scholar-gentry who owed their positions of authority to success in the civil service examination system rather than to any hereditary wealth. The growth of this meritocracy, combined with the development of a market economy in the countryside, led to a significant decline in the power wielded by well-off landowners – and the many responsibilities of actually working the land increasingly

**An 18th-century cotton wall hanging from northwest India depicts cows grazing and calves suckling in a paradisiacal landscape. The sacred status of cattle in Hinduism and the ban on eating their flesh probably originated because of the value of the living animals in agriculture. Cattle are still used for drawing ploughs in most of India, while their dung is used as a fertilizer and is the only cooking fuel for nine out of 10 rural households.**

devolved on those most directly concerned, the peasant families who made up by far the majority of the population.

The rural family household was the essential building block of the edifice that comprised the Chinese imperial state. The family unit was seen as reaching both back and forwards in time, encompassing generations of ancestors as well as unborn descendants; it was also linked by a great network of kin groups, lineages and clans to other family units occupying a sizable geographic area. Households were generally quite large, as the prosperity of each family was entirely dependent upon the number of able-bodied members it could put to work. On the other hand, because land was not inherited by the first-born exclusively, farms were divided into increasingly smaller parcels, and most were under two hectares.

The day-to-day life of rural families was far from easy or comfortable. In the 16th century a local historian living in Tancheng county in the province of Shandong, northeastern China, reported that although during a normal 12-year cycle it was usual to have equal periods of abundance and scarcity, his own immediate district was stricken by famine every year. When a newly appointed magistrate arrived in the county in 1670, local residents reported to him that their land had for a long time been "destitute and ravaged. For thirty years now our fields have lain under flood water or weeds; we still cannot bear to speak of the devastation." A succession of floods and locust plagues had repeatedly destroyed their harvests, and thousands of innocent citizens had been killed by gangs of marauding bandits. Despairingly, they went on to describe how, "On top of all this came the famine of 1665; and after the earthquake of 1668 not a single ear of grain was harvested, over half the people were dying of starvation, their homes were all destroyed and ten thousand men and women were crushed to death in the ruins."

In better times the land yielded sufficient quantities of food, but not without continuous work and attention. As soon as the winter snows had melted, the fields in Tancheng county where winter wheat was not growing were turned and spread with

**A 15th-century Italian illustration from a Hebrew law code shows the ritual slaughter of fowl and oxen. While many of the Jewish dietary laws laid out in the book of Leviticus may have evolved because they were hygienically beneficial, others have more complex origins. The pig, for example, was seen by early Middle Eastern pastoralists as a very different kind of animal from the sheep, goats, cattle and camels that constituted their herds: and it was possibly because the pig did not conform to the perceived norm that its flesh was outlawed.**

animal and human fertilizer. In early May this land was then ploughed and sown with millet and kaoliang; the earth was levelled with a wooden harrow and compacted with a roller or, in the absence of such equipment, by treading feet. As the crops grew, the fields were repeatedly weeded and the soil tamped down around the young shoots. The winter wheat was harvested in June, and the land where it had grown was ploughed and sown with soya beans. In August the millet and kaoliang were harvested and carried to the threshing ground, and in October the winter wheat was planted. Besides these staple crops, the farmers also grew a variety of vegetables such as turnips, cabbages, onions, celery, aubergines and garlic.

From the sale of their produce the farmers were obliged to pay taxes to both the local and central governments. In addition to taxes on land and on agricultural production, there were levies made on sales of a range of items including salt, livestock, wine and cotton goods, and the government claimed the right to buy produce from farmers at below cost price. All fit adult males were liable to be drafted for corvée labour – to build dykes, irrigation canals and other public works – or had instead to pay an extra tax in lieu of labour. Many households were also required to furnish a militiaman in times of emergency.

Unsophisticated though it may seem in comparison with modern Western agriculture, which made increasing use of technology and modern science from the 18th century onwards, it was this labour-intensive system based on the tilling of small plots by family units that enabled China to support its growing population into the 20th century. The very size of the workforce was one factor that encouraged such conservatism: where labour was cheap and plentiful, the introduction of machinery that saved on manpower was unnecessary, and indeed would merely have created new social problems by throwing thousands if not millions out of work.

Nevertheless, the growing population did place increasing pressure on agricultural resources. In 1740 the Emperor noted that "the population is constantly increasing, while the land does not become any more extensive", and he urged his subjects to cultivate every available plot of land "at the tops of the mountains or at the corners of the land". At the end of the first millennium China's population was around 100 million; by 1850 this number had risen to 450 million. In the 20th century it became apparent that the needs of the population could no longer be met without radical social reforms and technological development.

These became the urgent priorities of the Communist government following the establishment of the People's Republic of China in 1949, and they were acted upon with a speed and ruthlessness that was to turn Chinese society upside down. Land reforms that were completed in 1952 confiscated the holdings of landlords – two million of whom were killed in the process – and redistributed their land among the poor peasants. There then began a programme of collectivization, modelled on the experience of the Soviet Union in the 1920s, which involved the elimination of all private ownership of land. By 1957, 96 per cent of the rural population were incorporated into agricultural cooperatives in which all resources – equipment, animals and labour – were pooled for the common good. The cooperatives were in turn organized into self-sufficient administrative units known as communes, each comprising an average of 5,000 households.

Such sweeping changes could never have taken place had not the Chinese been accustomed for centuries to living in a hierarchical society in which obedience to one's elders and to those in authority was not just a social virtue but a way of life.

But in their attempts to substitute the state for the family as the focus of loyalty for all Chinese, the government inevitably created some unlooked-for disturbances. Reverence for the past had for countless centuries been a mainstay of social stability: evident in the chronicles of past dynasties painstakingly compiled by successive court historians and in the works of artists and writers who adhered to ancient traditions, its most basic expression was the ancestor worship of individual families. By seeking to eradicate all past models, the Communist leaders ensured that the birth of the new order, far from being a smooth transition, was accompanied by a great confusion and disorientation among the population.

In addition, the government's policies were often based on information that did not accord with reality. Following the establishment of the rural communes, the government came to believe, in line with the inflated production figures that had been supplied by ambitious commune leaders, that China had a massive surplus of grain, and it ordered a reduction in the area of land given over to rice growing. In fact, in

**In this 16th-century Japanese garden — a frozen encounter between resistant nature and the human urge to arrange and transform — the forms of mountains and rivers are tamed and reflected in a cultural mirror. Introduced from China during the seventh century AD, the art of creating these austere and compressed landscapes accorded well with the nature veneration of the ancient Japanese Shinto religion. After Zen Buddhism came to Japan in the 13th century, again from China, such gardens were created for Buddhist temples as ideal settings to foster meditation.**

the years between 1958 and 1961, China was suffering probably the worst famine in its history, causing the deaths of up to 30 million people.

Subsequent reforms, based more on a realistic assessment of China's needs than on ideological rigour, improved the lot of the long-suffering peasantry. Between 1965 and 1975 it is claimed that the area of irrigated farmland was increased by one third. In the 1980s individual peasant households once again became the basic unit of production, and they were permitted to sell surplus produce at rural markets. Overall, the food supply per person has risen since the Communists came to power.

At the beginning of the 1980s the government instituted a programme to reduce the rate of population growth. Families were officially forbidden to have more than one child, and lost their welfare benefits if they exceeded this limit. Although this represented a degree of state interference in private life that would have been unacceptable in the Western world, it was the logical response to the inescapable fact that China's arable land – comprising just 11 per cent of its total territory – could not support an ever-growing population.

Other countries around the world looked to China with keen interest, for in the 1980s there was a growing awareness that the problems of food supply were on a global scale. In 1990 it was estimated that 700 million people in the southern hemisphere were without adequate food – or in other words, for every affluent person in the north, another is close to starvation in the south.

Apart from limiting the size of families – and in most countries of Asia and Africa as well as in rural districts of China itself, effective birth control programmes have so far proved almost impossible to implement – only the application of advanced scientific procedures to crop growing could alleviate the scale of world hunger. In the 19th century the Western world solved its own food crisis by the mechanization of agriculture and the concomitant development of refrigeration, canning and improved transportation. In the late 20th century biotechnology – the range of disciplines associated with the genetic manipulation of living organisms – appeared to offer hopes of similar success. For example, crops with an in-built resistance to pests, and therefore not requiring the use of chemical pesticides, could greatly reduce the loss in production caused by weeds and disease, which at present destroy up to 40 per cent of the world's crops. But all such developments are vulnerable to both political and economic pressures. One variety of hybrid rice that is now grown in over a third of China's 33 million hectares of rice paddies has increased harvests by 25 per cent – but, determined to profit from this advance, China sold rights for the exclusive sale of the seed to two American companies, thereby limiting the number of poorer countries that could benefit from the development.

Some of these Chinese hybrid seeds have proved unsuitable for cultivation in Western countries because they require so much labour. The plants only make a few seeds, and without the availability of a large and cheap workforce to fertilize them by hand they offer scant profits. But this very obstacle demonstrates that the agriculture of China and other Asian countries, often viewed by outsiders in the past as backward and inefficient, has become a model of increasing relevance to other countries whose massive and rapidly growing populations appear to handicap all plans for a sustainable future. Labour intensive, economical of natural resources, with a strong emphasis on grains and vegetables, it is greatly productive in terms of yield per unit of land. Not least of its exemplary characteristics has been the degree of communal effort necessary for survival.

# EXTRACTING MINERAL WEALTH

In Africa, India and North America there are gold mines that extend so deep into the earth's crust that their walls are hot. Without air conditioning the miners who work them would die. When they return to the surface, raised in lifts which travel slowly so that their ears do not pop, their faces are unnaturally pale.

Humans are not built to burrow under the surface of the planet, but from the time they first began to fashion sharp tools and bright ornaments they have devised methods of extracting the riches embedded in the earth. Flint was quarried for making blades from around 40,000 BC. Copper was mined by the ancient Egyptians, and by Roman times the metal ores being extracted from both surface and underground sources included iron, gold, silver, lead, tin and zinc. In the 19th century the Industrial Revolution accelerated demand for coal and later for other fossil fuels such as petroleum and natural gas, all of which had been formed by the decomposition of prehistoric organisms.

The rate of material progress fuelled by the earth's minerals has been matched by the increasing impact of mining on the environment. The most conspicuous effects are those of surface mines – both those worked by manual labour, as shown in the photograph on the left of a gold mine in Brazil in the 1980s, and those exploited by mechanical excavators, the largest of which is capable of removing 180,000 cubic metres of coal per day. But the impact of underground mines is no less severe: they can cause subsidence, contaminate water sources and despoil the surrounding land. The furnaces for smelting silver ore from Roman mines in Spain, for example, consumed 500 million trees, clearing an area of 18,000 square kilometres. The safe disposal of radioactive waste from the mining and processing of uranium and plutonium for nuclear power now challenges the technology that, as shown on the following pages, has enabled humans to penetrate ever farther from their given habitat.

A clay tablet of the sixth century BC depicts Greek miners quarrying for clay. A ladder is propped against the side of the pit; the hanging amphora probably contained refreshment. On the far right, a 15th-century German woodcut shows a natural seepage of petroleum being collected in a jug. The oil was thought to cure numerous ailments: "He who pants with a cold or cough of long standing, let him rub the chest with it and it helps," declared a contemporary pamphlet.

In tunnels dug beneath the Japanese island of Sado, where gold was discovered around 1600, miners in a 19th-century woodcut by Hiroshige work by the light of burning wicks in vessels of oil.

# OPENING THE SEAMS

Although the majority of mines were dug and worked by muscle power until the 19th century, the impact of picks and shovels was by no means negligible. During the second millennium BC there were as many as 500 copper mines with timber-lined underground tunnels in the region of present-day Salzburg, while flint was mined in England from galleries supported by pillars of unworked rock. Along with human and animal muscles, the longest established and most enduring source of energy harnessed for surface mines was water, which was used by the Romans to break up banks of loose soil. The use of high-pressure water jets in the California gold rush from 1853 onwards actually served to alter the contours of the land, shaving soil from the hillsides and creating new hills of debris.

A German watercolour from around 1480 depicts an underground mayhem of pick-wielding miners. Because of the value of the iron ore they extracted, German miners were among the most privileged of medieval workers: many enjoyed exemption from taxes and military service.

In this silver mine in Lorraine depicted in a 16th-century German engraving, picks, hammers and wedges remain the miners' essential tools, but the removal of extracted ore has been streamlined by the use of man-powered winches and wheeled carts in vertical shafts and level, horizontal tunnels.

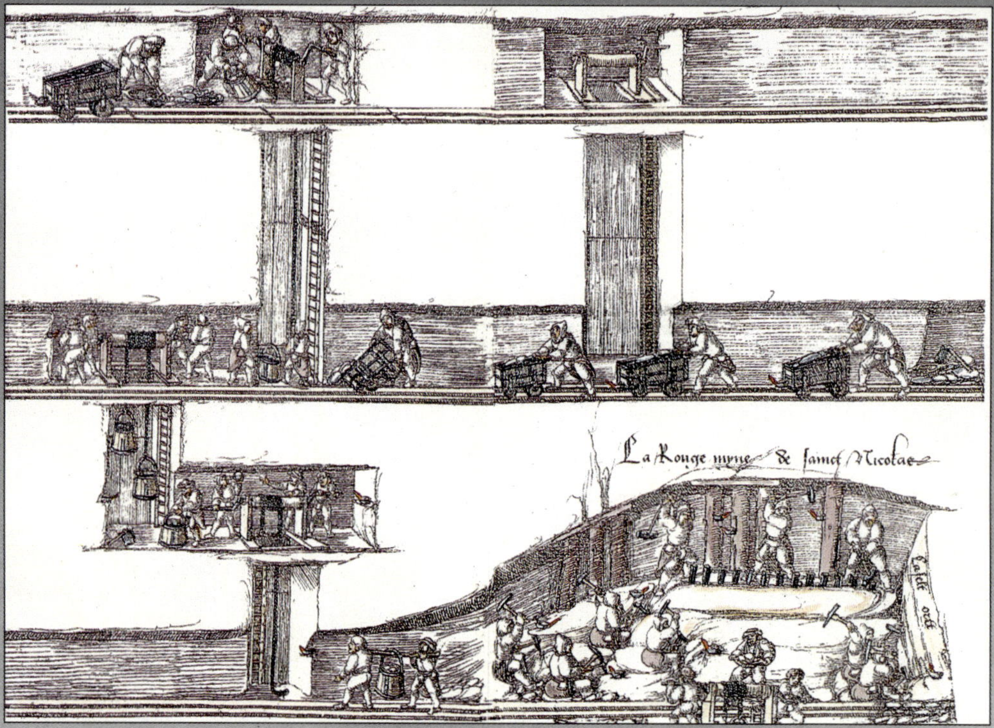

Woodcuts from a 16th-century treatise by the German mineralogist Agricola show drainage and ventilation systems. On the left, a man works the piston of a suction pump to raise water; on the right, men turn cranks to send air underground.

A steam engine at an early 19th-century pithead in England, feeding off the hemispherical boiler beneath the chimney stack, powers the winding gear that raised and lowered both miners and coal. On the right a loaded cart stands on a weighbridge.

# INCREASING THE YIELD

In medieval Europe, machines powered by humans, animals or streams were devised to bale water from flooded shafts and raise minerals from ever-lengthening tunnels. To mine iron ore – used to fashion tools, weapons and armour – hammers and wedges were supplemented by fires set to fracture a rock face and by gunpowder. Dwindling supplies of wood for the smelting of ores led to an increased use of coal, especially in England and Wales, whose output by 1700 amounted to five times that of the rest of the world. The development of steam power by the Scottish engineer James Watt from around 1765 provided a new source of energy with which to pump out flood water and, after the introduction of steam-driven rotary and piston drills in the early 19th century, to dig ever deeper.

A 17th-century Chinese wood-block print shows workers binding bamboo pipes that will be used to carry brine from underground salt mines. The pipes were also used as conduits for natural gas, which was tapped from salt mines as early as the fourth century. This gas was burnt to evaporate brine and also to provide heating and lighting for nearby buildings; in the 17th century whole towns were lit by its flames.

During the last century a twentyfold increase in demand for fossil fuels has been met by rapid advances in both the techniques and the scale of mining. The first oil well was drilled in Pennsylvania in 1859; oil is now extracted in every continent with the exception of Antarctica. Natural gas in strata under the land or sea was the world's single largest source of energy by 1970. The first nuclear power station, offering an alternative to the burning of exhaustible supplies of fossil fuels, was commissioned in Britain during 1956. New methods being developed to recover oil untapped by earlier technology include the use of chemicals that generate heat or radio frequency waves to make heavy oil less viscous, and the fracturing of impermeable rock by fluids which are pumped underground at high pressure.

A forest of wooden oil derricks crowd a hillside in Pennsylvania in 1865. Demand for oil – first used for lighting – accelerated after the development of the internal-combustion engine in the 1880s.

Five tugs tow a production platform from Stavanger in Norway to its station northeast of the Shetland Islands, where it started drilling for both oil and gas in 1975. Drill shafts and conductor pipes are housed in the concrete supports; undersea storage cells have a capacity of 1.1 million barrels of oil.

A worker at the U.K. centre for the reprocessing of spent nuclear fuel watches on a screen the operations he directs by means of the remote-control manipulator in his hand. The plant extracts reusable uranium and plutonium, and converts high-level radioactive wastes into solid glass blocks, which are sealed in stainless-steel containers.

# THE PLOUGH AND THE CROSS

4

Hollow-cheeked peasants, their bellies bloated from hunger, wandered aimlessly through the marketplace of Tournus in eastern France. Even if they had come with well-filled purses, or a basket of goods to barter, there was nothing to buy: the land lay in the grip of the worst famine that anyone could remember. Throughout the centuries of privation that followed the fall of the Roman empire, life was never easy for people striving, with poor tools and ill-fed animals, to scratch a living from the soil, but the year 1030 brought conditions that were nothing less than catastrophic. The weather all over Europe and far to the east turned cruel. For three years in succession the land was battered by violent, drenching storms. Seeds sown in the sodden ground were washed away; harvest time brought only a crop of weeds.

The chronicle of the Burgundian monk Radulfus Glaber told a grim tale: "After men had eaten beasts and birds, under the pressure of rampant famine, they began to eat carrion and things too horrible to mention. . . . It is terrible to relate the things which then befell mankind. Alas, a thing formerly little heard of happened: ravening hunger drove men to devour human flesh! Travellers were set upon by men stronger than themselves, and their dismembered flesh was cooked over fires and eaten."

Yet at the end of the year 1033, according to Radulfus, such suffering and hardship was miraculously ended. "At the millennium of the Lord's passion, the rains of the thunderclouds ceased in obedience to divine goodness and mercy. The sky began to smile. . . ." Famine and all its horrors, the chronicler assured his readers, were now a thing of the past, for the earth was once again fruitful.

This transition from dearth to fecundity was not as instantaneous or as universal as Radulfus implied. But his statement was in essence true: the time and place in which he lived saw, indeed, significant changes not only in the weather, but in the whole web of relationships between humankind and the natural world. From the 11th century until the start of the 14th, those who worked the land enjoyed a kinder climate, and the population rose at a rate far greater than at any time in the past thousand years. The pressure of extra mouths to feed spurred improvements in the techniques and technologies of farming, increasing the area under cultivation. And as woods were put to the axe and soil to the plough, Europeans learnt that the world was more malleable and its boundaries less finite than they had been taught to believe.

These developments were not just responses to environmental factors outside human control; they were intimately bound to the social institutions that dominated people's lives and moulded their perceptions. No medieval European lived beyond the influence of the Church and its teachings on relations between humans and the rest of God's creation. No lord or peasant lived outside systems of landownership and political authority that, while differing from region to region, invariably dictated how labour was performed and wealth distributed. But such institutions were not

**Labourers load a cart with newly cut logs in this detail from a 15th-century Flemish tapestry; behind them fellow woodcutters hew and saw, leaving raw stumps where once was woodland. In addition to being the only household fuel, wood was essential to every medieval industry: it fed iron-smelting and glass-blowing furnaces, was shaped into wine casks and vats, and was used to construct carts, ships, bridges, textile machinery and most houses. In the top right corner of the picture can be seen one of Europe's fast-expanding towns, in which the building of an average house used up 12 mature trees.**

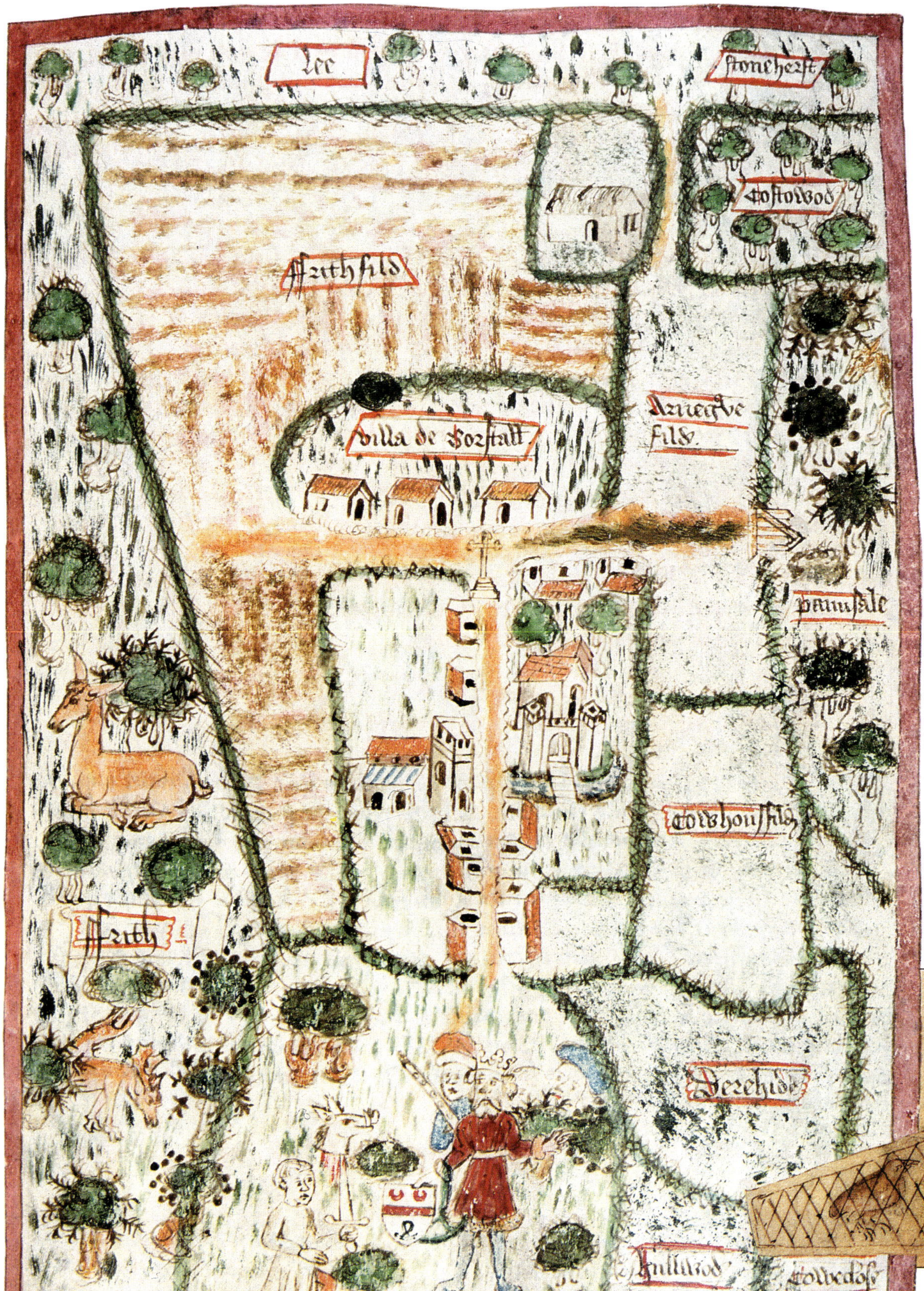
lee
stoneherst
ffrithfild
villa de Borstall
ffrith
Derehide

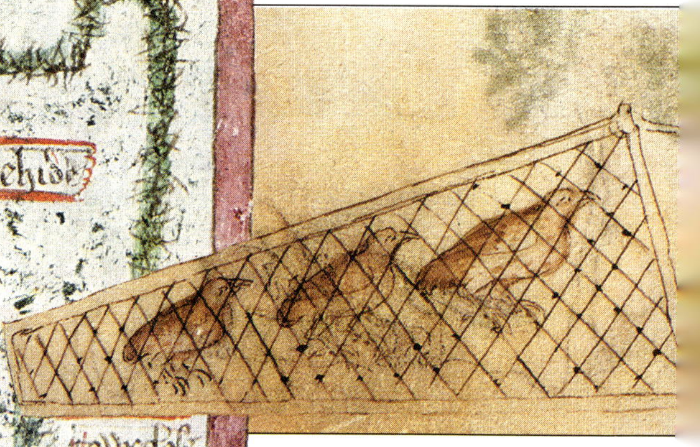

graven in stone, and as the fruits of affluence permeated through society their hold was loosened and a new order began to emerge. By the end of the 15th century an outward-looking and intellectually vibrant continent stood on the brink of an era of expansion that would spread European approaches to nature across the globe.

Europe's weather was, as always, subject to wide variation. The climate of a marsh or fen beside the North Sea had little resemblance to that of a high Alpine meadow, or to the baked Mediterranean plains. But in the northern and western parts of the continent especially, the 11th century saw markedly warmer and drier conditions, with milder winters and longer growing seasons. In some areas rainfall diminished enough to cause a drought, and settlements primarily dependent on rain for their water had to be abandoned; elsewhere, particularly in highland regions, villages deserted for a thousand years were now repopulated, thanks to an increasingly hospitable climate.

Travel became easier. In the Alps, where glaciers began to retreat and the snow line rose, mountain passes became accessible to travellers and traders for longer every year. Storms subsided in the North Sea, the Baltic and the Atlantic Ocean, allowing Norse seafarers to set out on voyages of exploration and conquest. They ventured beyond the known world to Iceland, Greenland and the easternmost tip of North America; they penetrated Russia and northwestern France through the European river systems; they sailed through the Strait of Gibraltar and appeared even in Byzantium.

The Norsemen did not take to their longboats simply through wanderlust. Like their fellow Europeans they felt the pressure of an increasing population. Some 90 per cent of Europe's people lived in small rural communities; these had formerly been self-sufficient but now, for the first time in centuries, there were more mouths to feed than the local fields could sustain. Meanwhile the continent's towns also expanded, attracting craftsmen and merchants who did not work the soil, but who still needed to eat. Townspeople made even greater demands on the countryside than did the rural peasantry, consuming some 15 to 25 per cent of all food produced in the hinterlands.

Responding to these demands, farmers who had fully exploited the best lands in their home territories were forced to cultivate less-promising ground, or to venture into altogether unknown country. In north and south alike, settlers moved tentatively into frontier lands; in mountainous regions from the Alps to the Pyrenees, shepherds moved their flocks across the high passes in search of new grazing grounds.

Every part of the continent had its own variations in landscape, soil quality and climatic conditions, and posed its own particular problems. Low-lying lands, from

**This map was made in the 1440s for Edmund Rede, the lord of Boarstall Manor, whose ancestor is pictured in the lower field receiving title to the land from King Edward I. Isolated and self-sufficient, Boarstall in Buckinghamshire, England, was a typical medieval village centred on its manor house and church. Crops in the common fields – cultivated on a three-year rotation cycle of winter wheat, spring oats and a year fallow – provided the main part of the inhabitants' diet. For meat, the villagers resorted to rearing pigs and hunting in adjacent woodland. The illustrations from a 14th-century Psalter *(below)* show the catching of partridges with a drop net and women using a ferret to drive a rabbit from its hole into a trap.**

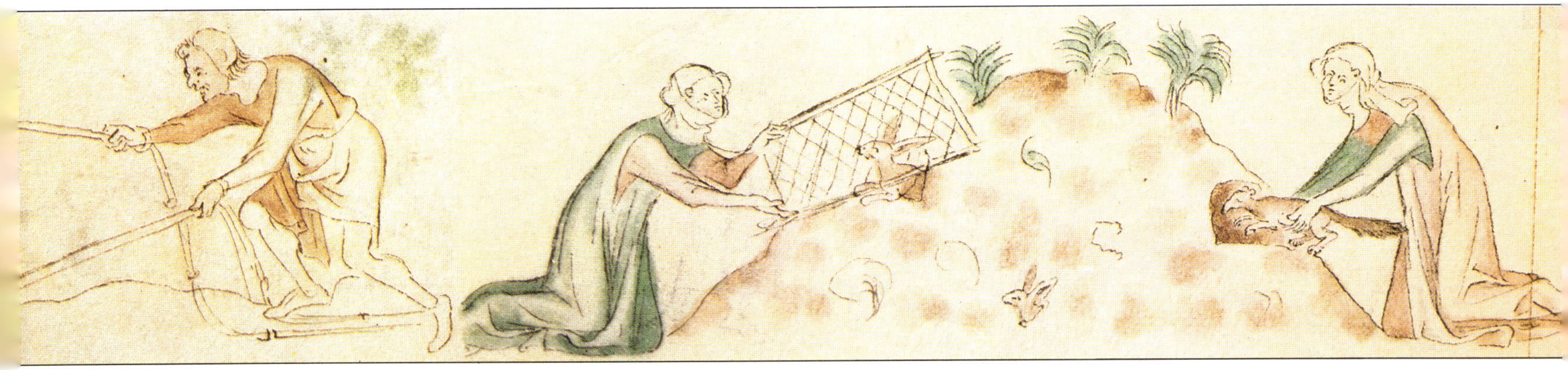

the East Anglian fens and the coast of Flanders to the malarial marshes of the Italian Veneto, demanded drainage before seed could be sown. In the arid expanses of southern Spain, where the Arab and Christian worlds met, settlers had to dig irrigation canals to make the land bear fruit. In the thickly forested north, German migrants pushed eastwards into the Slavic lands, hacking down vast tracts of woodland to make way for the cultivation of cereal crops, and simultaneously striving to sow seeds of Christian faith among the indigenous pagan population, still dedicated to the worship of their forest-dwelling gods.

The lords who owned land in these frontier regions swiftly realized that an influx of willing colonists, with strong backs and healthy appetites, could increase the value of their holdings. A Slavic chronicle told how the perspicacious count Adolf of Schauenberg brought new life to the underpopulated territories lying northeast of the lower reaches of the River Elbe. Making much of the land's unfulfilled potential, the Count "sent messengers into all parts, namely to Flanders, Holland, Utrecht, Westphalia and Frisia, proclaiming that whosoever were in straits for lack of fields should come with their families and receive a very good land — a spacious land, rich in crops, abounding in fish and flesh and exceeding good pasturage."

A further incentive to immigrants was the promise of far greater freedom and social mobility than their stay-at-home counterparts enjoyed. For the peasantry, especially in northern Europe, dwelt at the bottom of the rigid hierarchy of property and power relationships known as feudalism. This system — dominant in Germany, Flanders, England and northern and central France, but present also in Catalonia, Sicily and elsewhere in the south — was the product of centuries of political chaos, tribal invasions and social flux which had followed the collapse of the western portion of the Roman empire. It was rooted above all in a social contract between those who were — in both military and economic terms — the strong and the weak.

At the top of the social pyramid stood great warlords with large estates at their disposal. Since the introduction of the stirrup in around the eighth century, horses had played an increasing part in warfare, and battlefields had become the scene of clashes between charging adversaries, heavily armed and armoured, on thundering steeds. Only those affluent enough to buy and feed a horse, and to kit themselves out with the appropriate apparatus, could enter the military elite. And the basis of their income, almost without exception, was land, and any surplus wealth it yielded.

The warlords parcelled out their properties to fighting men of lesser fortunes, in return for their allegiance and support during times of strife. This emergent class of aristocratic warriors bestowed their protection upon the mass of humble noncombatants; in exchange, these peasant dependants grew the crops that sustained their masters' military adventures. By the 10th century the toilers in the field were not simply the social subordinates of the landed elite, but their legal inferiors as well, with their second-class status fixed by custom and enshrined in law. They had become manorial serfs, and were tied to the land they worked. They might possess a small patch of ground and some rudimentary tools, and perhaps also a pig or some fowls, but their daily and seasonal occupations, dwelling places, marriage plans and domestic arrangements were subject to the consent of the lord of the manor.

Outside the feudal lands, peasants — in theory — enjoyed greater freedom. In Spain, for instance, the territories of Leon and Castile, newly recaptured from the Moors by Christian forces, attracted Basque, Galician and Asturian colonists. Though dominated at times by local grandees, these smallholders kept their independence and never sank to the status of serfs. In northern Italy the citizens of the burgeoning towns owned much of the nearby countryside and gradually took control of the agricultural life within their own districts, setting quotas for grain, controlling cereal prices, banning or encouraging food exports as appropriate in periods of dearth or glut. In Pisa the authorities directed the local peasantry to plant a prescribed number of fruit trees annually, and required each rural household to cultivate bean rows and a cabbage patch. Their counterparts in Parma ruled that all fields be ploughed four times before a crop was sown, while the town of Modena designated the amount of land that was to be put to vines, and set an annual date for the grape harvest. Yet even among a free peasantry, autonomy was discouraged: the city of Milan passed a statute early in the 12th century requiring that peasants "show reverence to their lord".

The all-powerful institution that supported this social order, and laid down the rules of conduct for its members, was the Christian church. On all but the farthest fringes of the continent – Muslim Spain, the lands of the Balts and Slavs, or the forests and fjords of Scandinavia – the hegemony of Christianity was complete. And, as well as being a source of moral guidance and a dispenser of the sacraments that promised comfort in this world and salvation in the next, the Church was the wealthiest landowner in Europe. Its coffers bulged with rents, fees, tithes, bequests and charitable contributions. Its bejewelled reliquaries, embroidered vestments and illuminated gospels no longer provided easy pickings for heathen marauders: the raiders, or their descendants, had by now transformed themselves into baptized members of the congregation. And its wealth continued to grow as great princes, for the good of their souls, gifted vast estates with fat rent-rolls to their bishops, and the lords of rural manors donated tracts of land or subsidized the building of new religious houses.

Such offerings made sense in material as well as spiritual terms: any district stood to benefit by the presence of a thriving monastery in its midst. And from the 11th

**In a 14th-century Flemish manuscript illustration, a man bears a sack of corn towards a mill, where it will be ground into flour. Windmills of this type, based on earlier Arab and Roman models, made their first appearance in northern Europe in the late 12th century. The boxlike body, mounted on a sturdy oak post, could be turned into the wind by pulling on the tail pole protruding from its base.**

Miniatures from a breviary produced in Flanders in the late 15th century depict a range of agricultural tasks associated in northern Europe with particular months of the year. These include, from left to right: ploughing, spreading the land with animal manure and sowing in March; sheepshearing and reaping corn with sickles in July; knocking acorns to the ground to fatten the hogs in November, after which the animals will be slaughtered and their meat salted to preserve it through the winter; and spinning wool indoors in February while sheep huddle for warmth in the fold. In three of the four pictures the unchanging cycle of human activities is overseen and sanctified by the ever-present church.

century the monasteries themselves were becoming a new type of institution, with consequences that were to affect the natural world. Except within the still-pagan peripheries, the traditional task of making converts was no longer necessary; at the same time, the established monastic orders seemed to have lost the spiritual purity and rigour of their early Christian precursors. New members of the orders turned their minds to the Desert Fathers who, in the first centuries of the Christian era, had withdrawn into the wilderness, seeking salvation and enlightenment through physical labour, prayer and meditation. Alone or in small bands they set out to establish hermitages or small communities in wild, unpeopled places. By the beginning of the 12th century this revised approach to the cloistered life had become a significant force within the Church, and two new orders, the Carthusians and the Cistercians, had set up the first of many monasteries in remote regions: the former in the mountainous Dauphiné in southeastern France, the latter in Burgundy's Saône Valley, with its dense forests and unpromising sandy soil.

Pulling up dead and tangled roots, cutting down trees, clearing the ground of stones, coaxing reluctant earth to receive the plough, the members of the new ascetic

orders gradually transformed the landscapes to which they had retreated. The Cistercian monk Bernard, later to be canonized, spoke with pardonable pride of the work done by members of his own community at the mountain abbey of Clairvaux. Looking up from the cloister down in the valley, Bernard could see two hillsides: one slope made golden by the ripening grain that his brothers had planted, the other marked out with orderly rows of vines. "Each of them", observed Bernard, "offers to the eye a beautiful sight, and supplies a needful support for the inmates."

On the wooded summits of these hills, monks gathered dry branches to use for fuel, and cleared the ground to allow the native oak, lime and beech trees to grow without interference. Down below, closer to where the monastery itself stood, the River Aube ran through the valley in a course that the monks had purposely diverted in order to prevent dangerous flooding, to irrigate their fields, and to power a series of mills. Along its banks stood young orchards, heady with blossom and promising rich yields of fruit. To Bernard this carefully manipulated landscape was more than simply a tribute to the monks' good husbandry – it was a living sermon that was full of spiritual truth. "It greatly soothes weary minds, relieves anxieties and cares, helps souls who

Women sufferers from epilepsy or chorea are half-carried, half-dragged along, on a pilgrimage to a local church in this 17th-century copy of a lost original drawing by the 16th-century Flemish artist Pieter Brueghel the Elder. It was believed that dancing to bagpipes would rid victims of seizures for 12 months; one woman is held fast while a piper plays directly into her ear. Disorders of mind and body abounded in the Middle Ages, largely due to inadequate nutrition and to bread made from contaminated grain. Bread infected with the ergot fungus spread a fatal disease known as Saint Anthony's fire, which caused burning pains, hallucinations and gangrene.

seek the Lord greatly to devotion. . . . While I am charmed without by the sweet influence of the beauty of the country, I have not less delight within on reflecting on the mysteries which are hidden beneath it."

If the transformation of the wilderness was a spiritual mission for the brothers of Clairvaux and similar communities, it was also a hard-headed agricultural enterprise. In Flanders, for instance, the abbot Guillaume de Ryckel, recruiting peasants to help clear and work lands attached to the monastery of St. Trond, set out the ground rules for prospective tenants. For the first six years of their tenure new settlers were free to plant the crops they chose, in any quantity or order. From the seventh year, once farms were well established, the tenants were instructed to "observe the common custom of sowing, so that in one year they sow wheat or rye, in the second barley or oats or whatever spring crop is usually grown; and in the third year they will sow nothing." The Abbot's prescription reflected contemporary agricultural wisdom. To save the ground from exhaustion, all farmers appreciated the need to rotate their crops, to alter the foodstuffs grown, and at times to allow their fields to lie fallow.

In southern Europe the pattern of cultivation had not altered significantly since Roman times. Peasants worked small, enclosed fields, using a light plough suitable for drier, more friable soil. Farmers in the Mediterranean lands could not expect sufficient summer rain to give spring-planted crops a reasonable chance of survival, so the southern practice was generally to sow only one cereal crop in a year. Nor were hot, dry summers the only bar to high productivity. Lower grain yields meant less fodder to sustain livestock, and fewer farm animals meant less manure for fertilizer — the cycle was self-perpetuating. Nevertheless, in compensation for its relative poverty in cereal stuffs, the south had other riches: olives for oil, plentiful grapevines, and a long growing season for a cornucopia of vegetables and fruits known to few northerners.

North of a line drawn roughly from the Alps to the Loire Valley lay a colder, wetter region, scene of a slow but significant agricultural revolution which had begun in the eighth century and peaked in the 11th and 12th centuries. The farmers of the north adopted the so-called three-field system. They divided their land into three parts: a section for winter sowing of wheat or rye; a second field reserved for spring crops such as oats, beans and barley; a third piece of ground to lie fallow. Every year the functions of each field were rotated. The result was a virtual doubling of yields.

Whereas a light wooden plough had for centuries been adequate for farming the chalk lands of northern France or southern Britain, the clay soils of newly cleared valleys in the north needed heavy, wheeled iron ploughs of a type used in the Slavic lands, which turned the earth in furrows. The old light ploughs had needed only one draught animal to pull them – a donkey, mule, ox or, in times of desperation, a human being. The newcomers required the combined strength of six or eight oxen. Such resources were beyond the means of a single peasant and compelled farmers to pool equipment, so reinforcing the collectivist nature of village settlements. They

**A richly robed surgeon directs the activity of his pharmacy in this miniature from a 15th-century French manuscript. The assistant on the left grinds a medicine prepared from the ingredients in the jars behind him, while the youth on the right collects herbs from an enclosed garden. Beyond the arch lies the examination room. Qualified surgeons and physicians were capable of setting limbs, cauterizing and dressing wounds, and performing operations for hernia, gallstones, Caesarian section and plastic surgery, during which anaesthetics of mandrake or opiate-soaked sponges might be used. But they were able to offer little to sufferers of such incomprehensible diseases as malaria, scurvy, diphtheria or typhoid.**

also changed the face of the northern landscape. Because manoeuvring a hitched team within the tight space and awkward corners of a small, rectangular plot proved frustrating and time-consuming, there evolved by the 12th century a patchwork of large open fields, subdivided into long, narrow strips for efficient ploughing. An ox team could cover considerable ground before it had to change direction.

Few animals were generally kept on farms. Except in mountainous regions, meat and dairy products were only a small part of the diet for the bulk of the population, with most of this meat coming from the hunting of wild animals in nearby woodland. The only creatures bred exclusively for their meat were pigs, and these were virtually self-sufficient, allowed to run wild and forage among the oak and beech trees in the woods. Sheep were raised for wool; cattle were deemed as important for their hides as for their milk or meat; horses, donkeys, mules and oxen earnt their keep as modes of transport or beasts of burden. These animals were worked hard, but the most enlightened farmers recognized them as valuable resources, and treated them as well as their modest means permitted. "You must keep your plough beasts so that they have enough food to do their work, and that they be not too much overwrought when they come from the plough," advised Walter of Henley, bailiff in charge of the manors farmed by an English community of Dominican monks, "for you shall be put to too great an expense to replace them; besides, your tillage shall be behindhand."

Walter took pride in the smooth running of his farms, and felt he had gleaned enough wisdom to write a text on husbandry that other 13th-century agriculturalists would find helpful. The farms he managed grew wheat for bread, animal feed — probably the oats that in poorer places or times of dearth would nourish human beings too — and barley, a difficult crop but deemed worth the trouble for the sake of the beer and ale that were brewed from it. He might decide to put some of his land to inedible crops for industrial purposes, such as hemp for rope making or flax for cloth.

The warm, dry summers in this era of mild climate might even have tempted Walter to plant a vineyard: grapes were now grown successfully in his own southern English countryside, and on the Continent as far north as Flanders. But the heartland of the European wine trade lay to the east and south. Beside the well-trafficked rivers of central France, Burgundy, Lorraine, the Rhineland and the Moselle Valley, large quantities of wine were produced for export, a trade stimulated by the sacrificial role of wine in the Mass. In 1245 the Italian monk Fra Salimbene, visiting the vine lands of Auxerre, observed with a tinge of envy that "these people sow nothing, reap nothing, and gather nothing into their barns. They only need to send their wines to Paris on the nearby river, which goes straight there. The sale of wine in this city brings them a good profit which pays entirely for their food and clothing."

In the 14th century the proliferating vineyards of Bordeaux, ideally placed for the cross-Channel and Mediterranean maritime routes, became a powerful presence in the trade. The wine pressed from the grapes of the south made a fair exchange for the grain surpluses of the north, and both parties in these transactions prospered.

"Nose to the ground I crawl forward," ran a riddle of the time. "My nose is the grey enemy of the forest." To country folk the answer was easy: the iron tip of the plough. And as farming life moved to a more insistent rhythm, other simple but significant technological improvements took their place in the lives of Europe's farmers.

Wheeled implements in use by the 13th century included the spinning wheel, for turning flax or wool into yarn, and the wheelbarrow. Mills driven by streams — used in the Middle East and Egypt since at least the first century BC — advanced across the landscape. In the province of Picardy on the northeastern edge of France, manorial records noted that 40 were built between 850 and 1080, and more than 200 in the next 120 years. In England the Domesday Book — the great statistical survey compiled by William the Conqueror — counted some 6,000 mills in operation at the end of the 11th century. These devices could be turned to many uses: grinding cereal for bread making or brewing, pounding and processing cloth, powering the great hammer and bellows for a blacksmith's forge. And where regions lacked the benefit of rushing streams, the vanes of windmills rotated high above the flat terrain.

Following the development of a new form of collar harness with long leather traces and a flexible breast strap, horses came into their own as draught animals. A pair of horses harnessed in tandem in the new manner could pull three times as much as a team controlled with the traditional yoke, which tended to choke any beast pulling hard on a heavy load. The horse became an even more efficient helper after the introduction of nailed metal horseshoes, which provided better traction on difficult ground, and allowed horses to work longer, and pull harder, with less fatigue. A horse, which ate primarily grain, was more expensive to keep than an ox, which fed on cheap hay. But in northern Europe, with its relatively productive three-field system, the horse's increased efficiency outweighed its greater maintenance costs; in the south, where grain was scarcer, a single ox or mule still pulled the lighter ploughs.

The smith, hammering in nails to fix a shoe to a horse's hoof or mending an iron plough, became indispensible in every rural community, and the metals he worked were no longer luxuries or rarities in regions without mines. River barges and small seagoing ships were still the cheapest form of transport – and would be until the advent of the railway in the 19th century – but more and more goods were carried by carts along improved roads whose reach was extended by a proliferation of new bridges. Along their cobbled paving travelled silver from the Harz mountains, iron from the Slavic lands and Cornish tin to markets all over the continent.

The ring of metal axes resounded everywhere as peasants hacked down the trees of the forests. Woods had for centuries been used by peasants as a grazing place for their swine and as a source of domestic fuel. Their customary rights were enshrined in long-standing agreements with the lords of their manors: in exchange for the right to gather dead wood, a peasant woman might be expected to weave a cloth garment every year for her landlord; as a fee for letting her sow run free among the lord's oaks and beeches, she might owe a piglet from the animal's next litter. But the modest use of the forest for subsistence now gave way to a more thoroughgoing exploitation. Business-minded rustics collected wood far beyond their needs, and sold the surplus to residents of the rapidly expanding towns. There was also modest profit to be gained by the gathering and sale of squirrel skins, resin, charcoal, bark, wax and lime for a variety of crafts and industrial processes. But the most insistent demand was for timber itself – to construct new houses for the swelling population, to build merchant shipping, to make the barrels and chests that would carry the goods. For the first time,

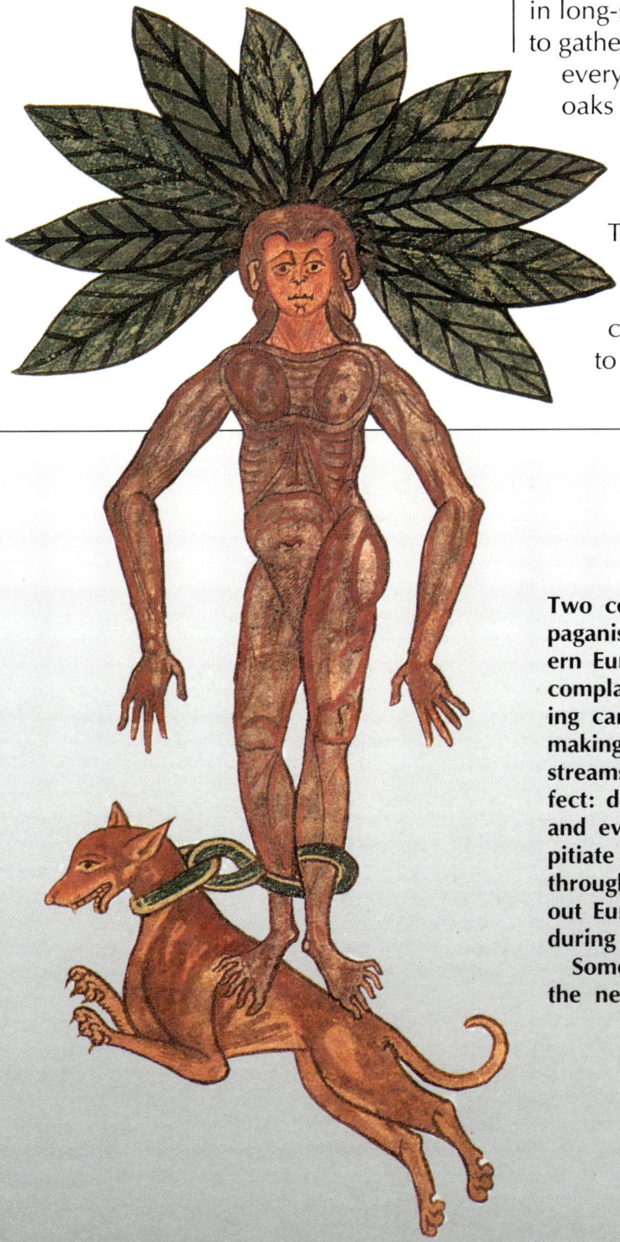

## ECHOES OF A PAGAN PAST

**Two centuries after Christianity replaced paganism as the official religion of Western Europe in 391, Saint Martin of Braga complained that peasants were still burning candles before rocks and plants and making offerings of bread to hallowed streams. His denunciations had little effect: deep-rooted popular beliefs in good and evil spirits, and in the need to propitiate the unpredictable forces of nature through pagan rituals, persisted throughout Europe at least until the Reformation during the 16th century.**

**Sometimes the old was subsumed by the new: legendary tales of folk heroes became identified with particular Christian saints, and certain objects associated with traditional ritual – from mistletoe to eggs – assumed a Christian significance. But many customs escaped altogether the fine net of Christian doctrine. The human-shaped root of the mandrake, for instance, shown on the left in an illustration from the 15th century, was believed to utter a shriek fatal to anyone who heard it when pulled from the ground, so a dog would be tied to the root to pull it up. Other such beliefs are illustrated on the following pages, some of which were the sources of superstitions that still survive.**

forests had to be planned and managed, with selective timber cutting and coppicing, and the tending of new plantations to replace what was taken away.

As trees crashed in greater numbers to the forest floor, old and new interests collided. The lords of the manors employed gamekeepers and bailiffs to ensure that no outsiders made free with their timber, deer, wild boar or any other fruits of the terrain. Legal documents bristled with disputes over rights of access and exploitation: feudal lords against their land-holding neighbours, peasants against each other, indigenous farmers against incoming monks, the monks of one abbey against those of another.

The expansion of land under cultivation also led to conflict between farmers and pastoralists — those communities with livelihoods based on sheep or cattle, whose open grazing grounds were now being eroded. Shepherds and cattle breeders in Iberia, the south of France and Italy were forced to move to new grazing grounds elsewhere, or to undertake the seasonal migration called transhumance — the bi-annual movement of livestock between mountain and lowland pastures, allowing intensively grazed ground time to recover and sprout new grasses before the herds or flocks returned. Even so, as the cloth trade burgeoned throughout Europe, the flocks grazing on the hillsides became part of a vast commercial and industrial mechanism: if the farmer in some districts drove out the shepherd, the balance was redressed elsewhere, where small cultivators saw their fields uprooted to allow large-scale rearing of sheep for the looms of Flanders and the fairs of Champagne. The monastic orders were among the first to enclose arable land for commercial sheep farming.

These changes did not happen overnight; most were half-perceived intrusions and gradual alterations to a pattern of life that must have seemed immutable. The mass of people lived in a world where wolves still howled in the wilderness; where tales of

**Sleeping men beneath the boughs of a yew in a 15th-century woodcut *(far left)* represent the funereal associations of the tree. Commonly found in graveyards, the yew was believed to bring dire misfortune or early death to any who slept or ate in its shade. In contrast, the hawthorn that bedecks the boat in the 16th-century Flemish watercolour on the left symbolized hope and fertility. Its blossom-laden branches were carried into houses on the first day of May as proof against maleficent spirits.**

heroes battling dragons, giants and enchantresses were as plausible as village gossip; where life spans were shortened by epidemics, malnutrition, dirt-borne infections and the perils of childbirth. Human life and the forces of nature remained intimately bound. Yet those who lived close to the land were not passive victims of their environment: they possessed practical knowledge – of the medicinal uses of plants, the behaviour of animals, the signs of changing weather – that helped them survive and prosper. They could read the world around them as fluently as scholars could a text.

Time was measured only loosely: there was little need or reason to compute a unit as small as a minute. The instruments that marked its passage – hourglasses, water clocks or sundials – were cumbersome or unreliable. Most of humanity tracked the progress of days by the disappearance or return of light; hours by the shifting position of the sun and the stars or, for clerics, by the prescribed sequence of daily prayers; weeks by the calls to Sunday Mass; seasons by the liturgical calendar, the changing weather and the cycle of agricultural tasks. The landscape itself was clock and calendar. "We were sitting under the elm," testified a 14th-century French peasant woman in a court case, "at the season when elms have put forth their leaves."

Space was even more nebulous a concept than time, and for most people the world was delineated by the distance they could cover in a day. Despite improved roads, travel for those whose lives and livelihoods demanded greater movement – itinerant pedlars, sailors, merchants, pilgrims, wandering monks, mercenaries, officials of the court – was slow and arduous. A laden ox cart might, with good going, cover five kilometres in an hour. A convoy of merchants or a nobleman with his retinue would be pleased to cover 35 kilometres between morning and evening; an unencumbered messenger riding a fast horse could, with luck, double that distance.

Despite these limitations, horizons were widening. By the 12th century the scholars, merchants and clerics of Christian Europe had come in contact with a society that was in many ways far more sophisticated than their own. The encounters were not

In a detail from Pieter Brueghel's painting *Peasant Wedding* the last sheaves of the harvest hang on a tavern wall. Preserved through the winter, they were believed to ensure a good crop the next year. In a carving from a 15th-century French misericord *(right)*, a smith shapes a horseshoe, whose supposed magical powers came both from its shape, resembling a crescent moon, and its material: the magnetic properties of iron and the arcane processes of smelting and forging caused the metal and those who worked it to be regarded with awe.

always peaceable: for many European Christians their first perception of the Islamic world came through the Crusades, a hundred years of religiously inspired military campaigns intended to wrest the Holy Land from Turkish and Arab control.

The Muslim communities of the eastern Mediterranean, North Africa and Spain had suffered far less from the political chaos and cultural disruption that had afflicted Christian Europe after the collapse of Rome. Since the dramatic spread of the Islamic faith in the eighth century, the Arabic-speaking community had provided a hospitable climate for the preservation and recovery of classical science and philosophy. In the Christian West the scientific and philosophical works of Aristotle, the geography of Ptolemy and Strabo, and the mathematical studies of Pythagoras had survived mainly as fragments, or through the distorting glass of commentaries upon long-lost texts. Within the Arab world, prolific translators of Greek originals provided a rich intellectual inheritance, and one that was rapidly expanded and advanced.

In their understanding of geography and astronomy, their techniques of map making, their navigational skills and their first-hand knowledge of remote regions, medieval Arabs seemed to live in an entirely different world from their Western contemporaries. They had assimilated the idea that the earth was spherical instead of flat, and divisible into lines of latitude and longitude. Their detailed navigational charts were considerably more accurate than any previous attempts had been, and Arab sailors were the first in the West to learn how to use a compass, possibly borrowing this device from the Chinese.

Armed with these skills, the Arabs built up a trading network from the Indian Ocean through the Persian Gulf, up the Mesopotamian waterways to Baghdad, and across the overland caravan routes to the Mediterranean Sea. The wares they carried made their way to European markets, where eyes widened at the sight of Chinese silks, Indian spices, tropical hardwoods and the improbably hairy fruit called the coconut. Less glamorous but more powerful in its impact was another Eastern import that

**A group of morris dancers entertain onlookers in this detail from *View of Richmond Palace*, an early 17th-century painting of the Flemish school. The dance evolved from the joint influences of ancient fertility rites and a Moorish dance known as the *Morisco*. The man dressed as a woman represents Mother Eve, while the hobbyhorse rider is descended from Saint George, the dragon slayer. It was believed that the stomping boots of the dancers and the hooves of the hobbyhorse had the power to rid the ground of the winter frost.**

Muslims bestowed on the Western world: in the 12th century they mastered the technique of making paper. Far cheaper to produce than parchment, paper turned books into tools instead of treasures and sowed the seeds for an intellectual renaissance.

Western scholars did more than just stand and wonder. They set to work studying Arabic, so they might translate this fund of new knowledge into Latin, the language held in common by all learned Christians. During the 12th century one scholar, Gerard of Cremona, translated an impressive collection of 71 works: 21 on medicine, 17 on geometry, 12 on astronomy, 11 on philosophy, four on geomancy, three on alchemy and three more on logic. In a memoir one of Gerard's students explained that his teacher, having learnt all he could in the Christian academies, travelled to Toledo, in the heart of Arabic Spain, where he studied the Arabic language. "In this way, combining both language and science, he passed on the Arabic literature in the manner of the wise man who, wandering through a green field, links up a crown of flowers, made not just from any, but from the prettiest."

Muslim and Christian minds met not only in the frontier land of Spain but along the trade routes controlled by Constantinople and Venice, and in Sicily, where the Norman king Roger II ruled a 12th-century court that mingled the culture, architecture and learning of Islam and the West. Crusaders travelling to or from the Holy Land by sea used Roger's realm as a staging post. They went home inspired by the learning of his court, by the Greek domes and Byzantine mosaics of his palaces, and by the magnificence of his country estates and gardens. An Italian visitor reported that "in order that none of the joys of land or water should be lacking to him" Roger had built a sanctuary for birds and beasts, and stocked its waters "with every kind of fish from diverse regions". Near his capital he created a park enclosed by walls that was "shaded with various trees and abounding with deer and goats and wild boar".

Northern princes took note, and copied. Robert of Artois, returning home to the northernmost part of France after a 13th-century Crusade, constructed a garden that compensated for the unattainable, balmy perfumed breezes of its Mediterranean prototypes by a dazzling array of fountains and ingenious moving statuary. Other travellers took cuttings of unfamiliar fruits, vegetables and edible flowers from Arab gardens: spinach, artichokes, oranges, lemons, apricots and saffron. And Christian travellers absorbed too the underlying philosophy of the Arab horticulturalists: the garden was a version of the oasis, an enclave where nature was tamed and transformed. In the north the less orderly world outside the garden's wall might be forest or scrub instead of desert sand dunes, but the principle was the same.

This new regard for the aesthetic beauties of nature testified to a more general change in European attitudes to the natural world. In the early centuries of the Christian era the teachings of the Church directed attention heavenwards: the next life, not this, was the proper focus for concern, and if plants or animals or the land itself had any significance at all, it was as a symbol of some higher spiritual truth. The pelican, for instance, which allegedly pierced its own breast to feed its young on the blood that gushed forth, was a symbol of Jesus Christ, sacrificing himself for humankind; the crab, with its sideways gait, embodied fraud or deceit. The earth was a textbook, written by God, interpreted by priests. Such beliefs persisted throughout the Middle Ages, but they were joined, and eventually supplanted, by an interest in the natural world for its own sake, and a growing enthusiasm for accurate observation and information.

The 13th-century Franciscan friar Bartholomaeus Anglicus was one of several encyclopedists who met this need with sweeping compendiums of knowledge that combined factual and fanciful in almost equal measure. In a passage on the crocodile, Bartholomaeus repeated the wisdom that was already old when it was retailed by the Roman author Pliny the Elder, noting that "if the crocodile findeth a man by the brim of the water or by the cliff, he slayeth him if he may, and then he weepeth upon him, and swalloweth him at the last." Bartholomaeus' description of Finland dwelt mainly on its inhabitants' propensity for witchcraft, especially in conjuring up winds for visiting sailors, while an inventory of human bodily organs assigned to each its commonly accepted function: "By the spleen we are moved to laugh, by the gall we are wroth, by the heart we are wise, by the brain we feel, by the liver we love."

Equal to the intellectual passion for acquiring knowledge of the world was the passion for arranging this information into orderly numerical categories. The composition of all animate and inanimate objects was based on the presence or absence of each of the four elements: earth, fire, air and water. The appearance, personality and health of every human being depended on the dominance of one of the four humours: the sanguine, the phlegmatic, the choleric and the melancholic. Good or bad fortune was intimately related to the influence of the 12 signs of the heavenly zodiac; spiritual salvation required the embrace of the seven cardinal virtues, and avoidance of the seven deadly sins. The theologian Albertus Magnus divided the world into seven climates, and showed how their peculiarities of weather and landscape affected the development of human beings and animal species. He explained, for instance, that the northern-dwelling races, such as the Goths, the Danes and the Slavs, had pale skins because of the climate they lived in: the frigid air constricted their blood flow, and turned their bodies white.

An age dominated by feudal and clerical hierarchies imposed the same imagery of rank and order on the universe. Heaven and earth were joined in a great chain of being, with God at the top. Human beings, sorted according to social status, took their places in the middle; then came animals, plants and inanimate objects. Just as men and women existed to serve God, and peasants to serve their lords, so lower forms of animal, vegetable and mineral were put on the earth to serve humankind.

But there were chinks in this ironclad system, as exemplified by the influential teachings of the Italian monk, Saint Francis of Assisi, who died in 1226. Saint Francis, founder of the order of wandering friars who came to be known as the Franciscans, insisted that all living creatures were equally worthy of kindness and respect. He attracted mockery by his practice of preaching to congregations of birds, on the grounds that they too deserved to hear the word of God, but ultimately came to be widely admired. He composed canticles of praise to Brother Sun, Sister Moon, Brothers Wind and Fire, and Sister-Mother Earth, "who sustains and governs us and produces fruits with colourful flowers and leaves!"

The dismissal of earthly pleasures as no more than a net of spiritual traps was giving way to a more sensual, if still reverent, celebration of the natural world. Where the fiercely ascetic old Desert Fathers might have sniffed brimstone on every gentle summer breeze, or seen the blooming rosebush only as a thorny snare of Satan, even a pope was now able to enjoy a ramble along the ancient Appian Way out of Rome towards Lake Nemi, remarking that "Nature, who is superior to any art, has made that road most delightful." On April 26, 1336, the Italian poet Petrarch with his brother and two servants climbed Mont Ventoux, about 45 kilometres northeast of Avignon — the first recorded ascent of a mountain purely for the enjoyment afforded by the view. "I could clearly see the Cévennes to the right," Petrarch wrote to a friend

A Flemish miniature from a late 15th-century edition of *The Romance of the Rose*, a narrative poem written two centuries earlier, shows many distinctive features of the medieval garden: elaborate trelliswork, raised flowerbeds, running water and a variety of fruit trees, some of them introduced to Europe after the Crusades. The rose, which in the poem symbolizes the fulfilment of love, also originated in the East. Mirroring the shape of a monastic cloister, the enclosed garden offered not only privacy and shelter from the workaday world, but an idealized setting in which the sensual pursuits of music and love acquired a religious significance.

on the same day, "and to the left the sea beyond Marseilles and Aigues-Mortes, all several days' journey distant. The Rhone itself lay under our eyes."

The lavish decoration of new churches and cathedrals reflected the shift in attitude. Saint Thomas Aquinas, prime philosopher and theologian of the age, declared that "works of art are only successful to the extent that they achieve a likeness of nature." Sculptors and woodcarvers concentrated on faithful copies of their favourite flora: snapdragons, ivy, cress and celandine, oak leaves, strawberries, ferns and broom, stone replicas of vine shoots twining round the doors. Fantastic creatures — the dragon or the basilisk, with its body a splice of bird and serpent — were juxtaposed with those faithfully copied from live models. Villard de Honnecourt, a celebrated 13th-century French architect, kept sketchbooks of his meticulous drawings: cat, swan, bear, lobster, dragonfly, and even a chained lion from a nobleman's private menagerie, its portrait labelled "Take notice that this was drawn from the life."

The Gothic architecture that reached its peak in the churches of 13th-century France, Flanders and England carried memories of the great forests that were now tamed and transformed. Long naves and pointed arches recalled the vistas through avenues of trees, and the shafts of light piercing tangled branches. The old pagan spirits of the northern forests – the guardians of trees and streams, the Green Man of local rites and legends – penetrated Christian sanctuaries as smiling grotesques, with leaves sprouting from their beards and hair, which peered from stone medallions above church porches or emerged from the carved and polished wood of choir stalls.

The building of the great Gothic cathedrals marked the climax of an era of optimism, intellectual energy and expansion in all sorts of human activity. But by the early 14th century the population increase that fuelled these developments was becoming a source of strain. Many parts of Europe stood close to saturation point; much of the countryside was overpopulated, with too many mouths for the land to feed. Soils in some areas were depleted, food prices high, and the wages for labour low.

As early as 1258, parts of rural England experienced the first in a sequence of serious food shortages, and by the first decade of the new century Germany was hit by famines nearly as grave as those recalled in such grim detail by Radulfus Glaber. And, to heap ashes upon the heads of those already suffering, the climate that had been so temperate since the millennium worsened. Longer and harsher winters caused crops to fail; remote northern or mountain settlements suffered from cut-off communications, languished and sometimes died out.

The gravest crisis of the age began in the 1320s, when the plague-carrying bacterium, *Pasteurella pestis*, emerged from rat burrows in the Gobi Desert in Mongolia, and – carried by the rat flea – entered China with its verminous host. From here the infection spread slowly along east-west trade routes, conveyed by the rodents inhabiting the caravanserais and delivered to towns and markets with the bundled merchandise. By 1346 the plague rats had reached the Crimea, on the Black Sea, from where they travelled by ship to European ports and then inland along roads and waterways. Ironically, the spread of the epidemic was facilitated by Europe's recent economic and technological advances: improvements in shipbuilding, resulting in vessels that were seaworthy all year, brought both better trade and the Black Death.

Among dense populations the germs were spread not only by fleas and rats but by the coughs and sneezes of people already infected. Even isolated settlements were at risk: one plague-carrying rat on a barge or one infected traveller could bring the pestilence, with its erupting sores, rapid fever and respiratory distress. Not all Europe was equally afflicted: the sparsely populated grain belt of Bohemia and Poland suffered less than the urban regions of northern Italy and Mediterranean France. But the overall toll was horrendous: at least a third of Europe's population died, either in the first onslaught at mid-century or in the recurrences of the 1360s and 1370s.

It would take at least 200 years before Europe's population returned to pre-plague levels. But, despite the devastation and the concurrent social tensions, Europe did not slide into stagnation. Economic and social recovery was slow but steady. With fewer mouths to feed, and less labour to help in the feeding, farmers cultivated only the best and most fertile land, which would provide the highest yields with the lowest expenditure of effort. More marginal, hard-won ground slipped back into wilderness. A shortage of workers meant earnings rose, and in many areas the reduced population found itself enjoying a considerably higher standard of living than ever before.

Moreover, those who survived or were born after the Black Death had freedoms and opportunities unknown to their forebears. The old feudal structures were crumbling: peasants were no longer serfs, owing labour and unquestioning obedience to their lords, but tenants, paying rents and free to move on to better opportunities

# A NOBLE PURSUIT

On Christmas Day in 1251, Henry III of England presided over a banquet including 630 deer, 200 boars and 2,100 partridges. This excess of meat was both a ritual expression of the King's wealth and symbolic evidence of his prowess as a hunter – for in medieval Europe, as in most of the world's preindustrial societies, a nobleman's quality was judged above all by his courage and skills in the chase.

Hunting served many practical purposes: it provided training for warfare, rid settlements of wolves and other predators, and contributed – through the creation of royal parks – to the conservation of woodland. But for many it was an end in itself, a specialized art form whose practice required an intimate knowledge of the behaviour of animals. This was especially the case when the hunter used trained animals to supplement his own weapons – a tradition that, as the pictures here and overleaf show, had long roots in many cultures. The Indian miniature *(right)*, painted around 1600, shows the Emperor Akbar hunting deer with a cheetah; on the left of the picture a blindfolded cheetah on a trolley waits to be released.

In a tomb painting dating from 1400 BC, an Egyptian with his wife and daughter hunts marsh birds from a papyrus raft. The hunter holds three decoy herons in his right hand, while his retriever cat has cleverly trapped three birds – one in its mouth, one in its front claws and a third pinned down by its tail.

A 19th-century Japanese ivory statuette depicts a falconer launching his bird towards its prey. Widely practised in Asia from around the eighth century BC, falconry was first introduced to medieval Europe by Crusaders and merchants returning from the Orient.

A huntsman pursues a deer *(above)* with the help of hawk and hounds in a 12th-century ironwork decoration from Sweden, where hawks were trained to peck out the eyes of their prey. On the left, a 16th-century tapestry depicts hounds savaging a boar while their master, the Holy Roman emperor Maximilian I, prepares to deliver the fatal thrust with a sword specifically designed for the sport.

In a 10th-century Chinese drawing, hunters returning from the Asian steppes share the saddle with their Saluki hounds. The Saluki, a breed domesticated as early as 7000 BC, was used in hunting in ancient Mesopotamia and Egypt.

elsewhere. Towns, though no longer bursting at the seams after the depredations of the plague, were becoming the true centres of economic life and power, not only in the southerly Mediterranean lands but also in the north. The lord of the rural manor now had a potent rival in the person of the urban merchant, who preferred trading missions to Crusades, felt more at home in the counting house than on the battlefield, and showed his loyalty to his monarch by making loans instead of feudal vows.

Nor was the advance of science and technology slowed by the years of plague. By the early 1400s new Latin translations of classical authors – Ptolemy on geography, Euclid on geometry and Vitruvius on architecture – made these writers and their ideas more accessible, and thus far more influential. Gunpowder, acquired from the Chinese through the agency of the Arab world, transformed the logistics and machinery of warfare. The mining industry, responding to the increased use of metal coinage and the growing demand for iron tools, became more productive through various new devices: suction pumps, cogwheels and systems of gears, powered by horses or by water. Even the measurement of time had progressed: the first weight-driven clock had been installed in Milan in 1335, and by the beginning of the 15th century it was a poor city that did not boast a similar dial on the tower of its cathedral or town hall.

If the world had changed, so too had the ways of looking at it. Nowhere was this clearer than in the transformation of the visual arts. Northern painters of the Flemish

school scrupulously reproduced the hunting parks of local gentry, and both known and imaginary landscapes, as the background to religious and secular scenes. In the brilliantly detailed illustrations to the book of hours created for Jean, the duke of Berry, the artists depicted the turning year through the actions of men and women within a landscape bearing all the marks of human intervention and control. And in Italy painters experimented with new theories of proportion to reproduce the human form in a way that was not only mathematically accurate but aesthetically pleasing.

Subject matter developed accordingly. The religious scenes that still dominated artistic production were no longer flat, impersonal images, but dramas played out by recognizable human beings, who wept, bled, suffered or rejoiced as they might have done in life. To further these efforts to imitate reality, artists such as Brunelleschi developed the use of linear perspective — the technique for creating a picture where the size, distance and relative positions of all figures and objects faithfully imitated the spatial relationships that might be seen within a similar slice of the real world, as viewed by someone just outside the frame, peering in as through a window.

While painters studied methods for making line, light and shadow into plausible copies of the subjects they represented, a complementary breakthrough was taking place during the 1440s in a workshop in the German city of Mainz, where Johann Gutenberg and his colleagues developed a method of printing books by means of movable metal type. The art of printing was not a new discovery: the Chinese had been reproducing texts on a form of mechanical press since the ninth century, but the technique had apparently never travelled farther west than Turkestan in central Asia. Once Gutenberg introduced this mode of mechanical reproduction to the West, however, it spread more swiftly, and more beneficently, than the plague. By 1474 books were coming off presses in Paris, Lyons, Venice, Naples and the Polish city of Cracow. By 1500, print shops in 236 European towns and cities had turned out an estimated 20 million volumes — at a time when the whole population of the continent comprised no more than 70 million souls.

Linear perspective and movable type together aided a revolution in the production and transmission of information. New ideas were accessible to those able to master the skill of deciphering letters; illustrations, maps and diagrams could be easily and clearly reproduced. Intellectual stimulation was no longer the preserve of clerics in monasteries and rich patrons in palaces, poring over rare, hand-copied texts. Facts, fresh discoveries and contentious theories — on medicine, mathematics, geography, chemistry, botany and astronomy — flew back and forth among Europe's scholars.

Some of the new texts posed dangerous challenges, and the guardians of religious orthodoxy would strive to consign them, and their authors, to the flames. The conservatives' fears were understandable: Christians could now read the words of the Bible themselves and make their own interpretations of religious truth. But the flow of new ideas was unstoppable. Anyone able to read — and the numbers rose at a prodigious rate — had no reason to be a mere recipient of someone else's view of the world. All of Creation was open to question, and every question was soon to be asked. The printed word, that most arbitrary set of abstract symbols, had opened the floodgates. The book of nature was about to be rewritten.

**A painted sign from a Venetian workshop shows men at work on six vessels in the Arsenal, the greatest shipbuilding complex of medieval Europe, which took its name from an Arabic word for "house of industry". The Arsenal supplied Venice with war galleys and a merchant fleet of over 3,000 ships; the latter included light inshore caravels and square-rigged cogs sturdy enough to sail into international waters. The growth of sea trade in Europe, dominated by Venice and Genoa, stimulated improvements in ship design that made possible the voyages of discovery to the New World.**

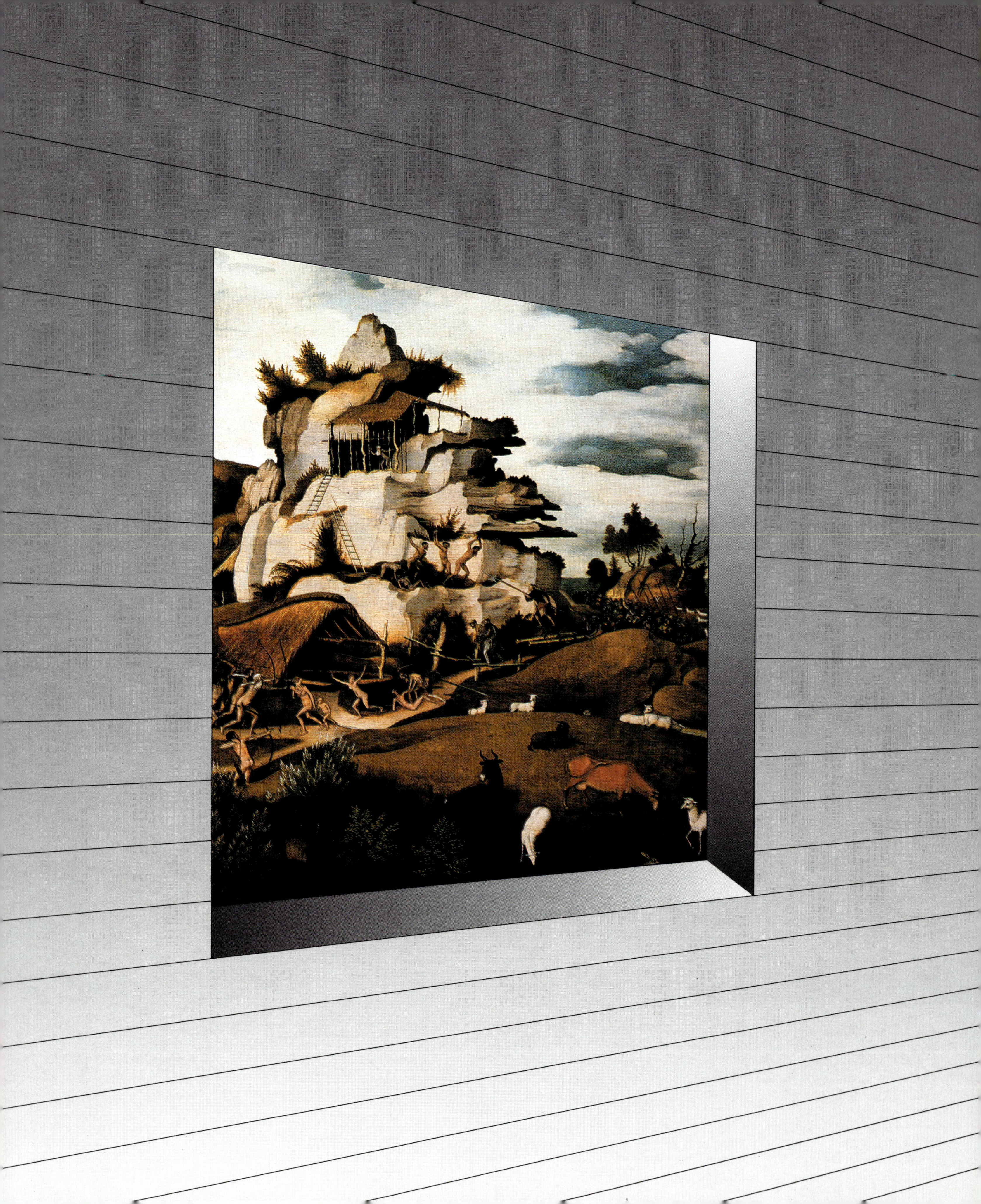

# EXPANSION AND EXCHANGE

5

In March 1519 runners brought to the Aztec emperor of Mexico tidings of new and outlandish invaders who had arrived on the shores of his domain. "The strangers' bodies are completely covered," reported the messengers, "so that only their faces can be seen. Their skin is white, as if it were made of lime." And what was even stranger still: "Their deer carry them on their backs wherever they go. Those deer, our lord, are as tall as the roof of a house."

For the emperor Montezuma there were only two possible explanations: either his messengers were liars, or his territory was being invaded by supernatural beings. In fact neither was correct – but the Aztecs had never seen white-skinned men before, nor had they seen horses. And the paralysing fear that was induced by the appearance of these wondrous new species so sapped the Aztecs' will to resist that within just two years their massed armies had been vanquished by a force numbering just a few hundred Spanish conquistadors.

The Spaniards had other advantages besides their appearance. Most obviously, they were armed with swords of cold steel and cannons that were capable of inflicting damage on a scale immeasurably greater than the Aztecs' own spears and arrows. "A thing like a ball of stone comes out of its entrails," reported the Aztec envoys of this terrible weapon. "It comes out shooting sparks and raining fire." Over the following centuries, however, the chief means by which European invaders dominated both North and South America was not military hardware but a more subtle, though equally effective, biological arsenal of diseases, animals and plants. Isolated from the continents of Europe and Asia for thousands of years, the peoples of the Americas lacked immunity to the invaders' ailments. And in the wake of the Spaniards' horses came a menagerie of other animals wholly new to America – cattle, pigs, sheep and goats among others – whose voracious appetites forced many indigenous plants to the point of extinction. Finally, like a vanguard of skirmishers in a military campaign, Europe's weeds preceded the inexorable advance of Europe's wheat, barley and other crops, which helped fuel a rapid increase in the numbers of the new animals and people alike. By 1800, much of what had at first been to the Europeans only an alien wilderness had been transformed into a virtual home from home.

The pace of this ecological invasion, although it eventually came to develop a momentum of its own, was forced by very human factors, not least the Europeans' assumption that the riches of the natural world were theirs to exploit by right, and that success could be measured in terms of economic profit. And sea power gave Europe the opportunity to impose a similar form of biological colonization wherever they made landfall. In some regions the graft did not take; but wherever climatic conditions permitted European flora and fauna to flourish and the land was not already densely populated by resistant peoples, the story of the American continent was

**A detail from a canvas painted by the Dutch artist Jan Mostaert in the 1540s shows American Indians on a craggy outcrop hurling down stones on an advancing column of Spanish conquistadors. The painting may have been based on an eyewitness account of a 1540 battle fought in present-day Arizona between Spanish soldiers – who were tracking a rumour of gold – and Zuñi Indians. Although Mostaert captured the true savagery of the conflict between New World inhabitants and Old World invaders, his vision of a pastoral landscape where sheep and cattle graze calmly in a well-wooded field was more in accord with European expectations and artistic convention than with reality.**

TYPVS ORBIS TERRAR
SEPTENTRIO.
AMERICA SIVE INDIA NOVA. Ao 1492. a Christophoro Colombo nomine regis Castellæ primum detecta.
CIRCVLVS ARCTICVS.
TROPICVS CANCRI
CIRCVLVS AEQVINOCTIALIS
TROPICVS CAPRICORNI.
CIRCVLVS ANTARCTICVS.
MAR DEL ZVR
MAR DEL NORT
EL MAR PACIFICO
OCEANVS AETHIOPICVS.
EVROPA
AFRICA.
Tartaria
Terra del Fuego
Hanc continentem Australem, nonnulli Magellanicam regionem ab eius inuentore nuncupant.
TERRA AVSTRALIS NONDVM
MERIDIES.
QVID EI POTEST VIDERI MAGNVM IN REBVS HVMANIS, CVI AETE
OMNIS, TOTIVSQVE MVNDI NOTA SIT MAGNITVDO. CI
Cum privilegio.
Franciscus Hogenbergus sculpsit

successfully replayed. There were some indigenous plants and animals which travelled in the opposite direction, it is true, but these too made their contribution to the long-term advantage of Europe. They supplemented Europe's existing diet, and they fuelled intellectual and scientific advances that would ultimately enable the colonial powers to extend their reach still farther.

The advantages that the Europeans were able to exploit so successfully were the product of millenniums of evolution. Some 200 million years ago, the slow grinding of the great rock plates that make up the earth's outer crust began to tear apart what was then the planet's single landmass, dividing it into separate continents in which plant and animal life evolved in distinctive ways. The human inhabitants of these new continents, too, developed their technological abilities at different paces. As a result, when the Europeans at last came face to face with their long-lost cousins, the parties scarcely recognized the kinship between them. The world had suddenly become larger and more various than anyone had previously supposed it to be, and challenged every convention by which it had been understood.

Captured in the grid of a map from the first atlas of the whole world, published in Antwerp in 1570, the recently discovered continents of North and South America are suspended with Europe, Asia and Africa above the hypothetical landmass of Australia. Following the transatlantic voyages of Christopher Columbus in the 1490s, peoples on both sides of the ocean were forced to revise their understanding of the world. To the Indians of southwest America who painted on a rock face the Spanish cavalcade shown above, even the horses that bore the invaders were utterly new.

The first humans to reach America probably arrived sometime before 13,000 BC, during a period when the continent was linked to present-day Siberia by a land bridge laid bare by falling sea levels during the last ice age. To the south of the slowly shrinking ice fields they discovered a vast new territory, free from any human rivals they might have to compete with and abundantly provided with the big game they had become so efficient at killing with their spears and flint arrowheads. These peoples' numbers increased as they spread southwards; by around 10,000 BC they had established themselves everywhere in the continent from the edge of the Arctic to the distant tip of South America. But by that time the melting glaciers of the dying ice age had raised sea levels and closed off the bridge from Siberia. Just as happened to their fellow pioneers in Australia, who were also trapped on their continent by rising sea levels, the new Americans were left isolated from other humans and would develop in their own way.

They made the most of the natural resources they discovered. America had existed as a separate biosphere for millions of years, and the newcomers remained largely untroubled by the irksome parasites and predators that had evolved alongside their primate ancestors in the Old World. Similarly, the animals of this continent had never needed to evolve a defence against primate rapacity, and would never now get the chance to do so, for within only a short time most of the larger species — including mammoths and the original American horses — had become extinct, at least in part because of the large-scale hunting of the human inhabitants. But the Americans managed to adapt their weapons for catching smaller game and thrived on what they could gather from the fertile land. Around 5,000 years ago those living in the most crowded part of the continent — an area approximating to that of modern Mexico — began to cultivate wild maize. This crop was improved over the course of generations to become America's most important native food plant, and was supplemented by growing squashes and beans, which were both richer in protein. By 1500 BC village-sized settlements were common; by 500 AD Mexico supported some notable cities, along with the complex, organized society that went with them. Though they failed to put the wheel to productive use and had no knowledge of iron, they were learning the skills of metalworking, especially in gold and silver. The human invasion of the Americas could be reckoned a success.

The far more numerous humans who were left behind in the Old World, however, had done even better. To begin with, they adopted agriculture thousands of years before their American relatives did, and whereas the Mexican protofarmers took many centuries to increase their maize yields to a level where they were capable of supporting a civilization, the wild wheat that the Eurasian cultivators began with was already heavy with plump and nutritious grain.

At least as importantly, the domestication of key animal species was a skill that was late in coming. At the time the first humans had entered America, only dogs had been successfully tamed, and only dogs accompanied them. During later millenniums Old World humans domesticated a vast roll call of species that included cattle, sheep, goats, horses, pigs, buffaloes, reindeer and even honeybees. But when the Americans, too, began to acquire the technique, they found that they had a shortage of suitable subjects. The dearth may have been at least in part due to their early success as hunters. At any rate, against the formidable array of species that had been brought into the service of Old World humanity, they could muster only the llama, the guinea pig and a few kinds of poultry.

The wider variety of animals they had learnt to make best use of gave the Old World humans immense advantages, of which the availability of ready supplies of meat, hides and fertilizing dung were only the most obvious. Ploughs pulled by horses and oxen could cultivate land that was beyond the strength of humans to dig, and pastoralists were able to live off their herds in marginal country that would support cattle but not crops. Some Old World humans even took their adaptation to domestic animals a step further: the stomachs of Europeans in particular had learnt to thrive throughout adulthood on a diet that included milk; few Asians and no native Americans were able to tolerate milk after infancy.

The Old Worlders also learnt how to cope with the disadvantages that came from the propinquity of man and beast — and here they would find what proved their most powerful weapon against their distant cousins. Most of their domesticated animals carried microorganisms and parasites such as fleas that could cross the barrier between species and afflict humans, usually much more severely than they did their original hosts. But, given enough time, the survivors evolved at least partial immunity to many such infections. Besides, as Old World cities developed, with many centuries' head start over the nascent civilizations of America, the most dangerous of the infections — smallpox, for example — became endemic in their crowded populations.

**A vision of order stamped upon an alien wilderness, this engraving after a drawing by Peter Gordon shows the uniform building plots of the young settlement of Savannah in the American colony of Georgia as they stood at the end of March, 1734. The town had been founded in the previous year by a British Member of Parliament, James Oglethorpe, whose tent is pitched between the riverbank and the first houses. The colonists, whose land-holdings were increased from 200 to 800 hectares each in 1740, were encouraged to grow wines, spices and silk to relieve Britain's dependency on more expensive sources. In 1819 Savannah was the embarkation point for the first steamship to make a crossing of the Atlantic.**

Disease still killed many thousands, but usually in childhood; adults were left with an armoury of antibodies. The people of the New World, who had never had to deal with anything like the Old World's pool of lethal pathogens, had no such defences. In comparison with the disease-hardened men and women of the earth's central landmass, they were biological innocents.

For several thousand years, America remained isolated by the quarantine of distance. But in the 15th century the peoples of Europe, led by the Spanish and the Portuguese, steadily acquired a knowledge of the pattern of the transoceanic winds, without which no-one could hope to cross an ocean and return alive. And they were equipped not only with ships that were capable of carrying them round the world, but with the feelings of cultural superiority that would-be conquerors needed just as much as they did weapons and charts.

By the 1440s the Portuguese were settling the uninhabited isles of the Azores and Madeira, transporting their own plants, animals and slaves there from their homeland and raising sugar cane which they traded to supply Europe's insatiable sweet tooth.

Farther afield, their ventures in the Indian Ocean brought in return profits that more than compensated Portuguese mariners for the hard apprenticeship they had undergone to learn the ways of the ocean winds. Portugal's Iberian neighbour Spain got entangled in an utterly one-sided genocidal war against the Stone-Age Guanche people who lived on the Canary Islands. And when the Genoa-born mariner Christopher Columbus began begging and wheedling his way round the courts of Europe to raise funds to finance a voyage westwards from the Canaries, it was the prospective riches to be gained from a new route to China that eventually persuaded Ferdinand and Isabella of Spain to back his venture.

With his 90 men and three small ships, Columbus made an unspectacular landfall in October 1492 following an unexciting voyage which had lasted just 33 days, an almost absurdly easy cruise when compared to the wide, pelagic swoops of Portugal's India-bound mariners. Columbus continued to believe up until his death in 1506 that he had found his longed-for route to China. But, of course, not only was he not in the East Indies, he was nowhere near them – his ships had touched instead on the Bahamas, and he had discovered not a back-door route to the other end of Eurasia but an entirely new continent.

The Spaniards did not delay in exploiting the breach that Columbus' voyage had opened. By the beginning of the 16th century a horde of adventurers had followed in Columbus' wake, and the Iberian appropriation of the island of Hispaniola was nearly complete. For the first few years the Spaniards contented themselves with little more than pillage, seizing gold and slaves to take home with them. Soon, however, they settled into serious exploitation, growing sugar in a successful attempt to match the lucrative production of Madeira and the Azores. The pigs, cattle and even horses Columbus had brought were breeding with enthusiasm, rooting out and trampling down what remained of the original ecology. Many of the islanders were killed;

## THE NATURALISTS' RECORD

Diamond python, painted in Australia around 1790.

**The contents of the pouch held by the Swedish naturalist Carolus Linnaeus in this portrait *(right)* include pens, brushes and a microscope. No less important than guns and ploughs in the European appropriation of the Americas and Australia, these tools were the instruments by which new-found flora and fauna were observed, recorded and caught in the expanding web of Western science.**

**The classification system for plants devised by Linnaeus enabled botanists to place discoveries into named categories, and so establish their relationship to other species. Other naturalists – such as those on the following pages – also contributed to a more rational and objective understanding of the natural world by their interpretation of the data they collected.**

most of the rest became slaves on the sugar plantations their new masters created. To eke out the native labour force, the first African slaves were hauled across the Atlantic in 1505. And in 1518 – with fatal consequences – another immigrant arrived: the smallpox virus. Barely a thousand of the original Amerindian population of Hispaniola survived, according to one witness. A scant generation after Columbus' lookouts first saw the island loom above the Atlantic swell, its original people, animals and plants had been virtually annihilated – and rapidly replaced.

On the offshore islands smallpox merely helped finish off a job that the Spaniards already had well in hand. On the mainland, where settlement had begun in 1509, it was decisive. A year after the tiny band of conquistadors led by Hernán Cortés launched their against-the-odds invasion of the huge Aztec empire, the sickness crossed over along with reinforcements from Hispaniola. The extra humans were not nearly enough on their own to turn the balance, but the virus that one of them was carrying more than made up for their lack of numbers. Within a few months a raging smallpox epidemic, to which Cortés and his men were almost all immune, had killed off perhaps half the native population. The death rate was highest among the young adults – a grim commonplace in any encounter between an infection and a human group with no established immunities against it – and those who survived were left utterly demoralized. Victims of invaders who appeared to have divine or diabolic power on their side, the Mexicans soon found themselves reduced to the status of serfs of the "viceroyalty of New Spain".

The disease then spread southwards, ravaging the Inca dominion of Peru and thus opening the way for another improbable victory a decade later, when Francisco Pizarro added the viceroyalty of Peru to the Spanish crown. It ran still

**In a 1737 painting *(left)*, Linnaeus is dressed in the traditional garb of Lapland, where he travelled to study plants in 1732. A notebook sketch *(above)* shows the Greek mythological figure Andromeda, who was chained to a rock, and the flowering marsh plant that he named after her.**

**Lobster krill, painted off the coast of Patagonia in South America in 1769.**

further, burning like a forest fire in a drought, the spark of new outbreaks of the sickness often borne by refugees who carried in their own bodies the death they hoped they had left behind. And following in the wake of the pestilence came the empire of Spain, relentlessly expanding down the length of South America, with a plethora of other lethal illnesses – including measles, mumps and diphtheria – seemingly ready at hand for those whom smallpox failed to exterminate.

Infectious disease was the single most powerful weapon the Europeans had against indigenous peoples. Against indigenous ecologies, their plants and livestock were just as effective, generally without their owners' active intervention. The most startling example was in the vast plain of the pampas, the grasslands that extend across the area of modern Uruguay, northern Argentina and southern Brazil. The Spanish abandoned a failed settlement in the region early in the 16th century, but though the colonists went home, they did not take their animals with them. They ran wild and prospered, for they had no competition: there had been no grass-eating ruminants in the region for thousands of years. When the city of Buenos Aires was successfully founded at the second attempt in 1580, there were already substantial herds of feral cattle, and from then on the bovine population growth was explosive. The local grasses, which were not adapted to withstand steady grazing, were soon wiped out. But the numbers of cattle – there were plenty of wild horses, too – continued to rise, because the native grasses that died were quickly replaced by their European analogues, which had evolved over millions of years to thrive despite heavy cropping. In effect, between them the partnership of plant and animal, with no more human intervention than was necessary to transport them across the ocean, Europeanized an immense tract of South America.

The colonization of the northern part of the American continent had to wait until its English, French and Dutch protagonists had developed enough sea power to enable

**A 1795 cartoon mocks the rise to social prominence of the British naturalist Sir Joseph Banks by showing him as a butterfly evolved from a grub. On the opposite page, three pictures show the stages by which some of Banks's discoveries passed into common knowledge: the plant collected in Botany Bay, Australia, and named after Banks; an on-the-spot sketch of the same plant executed on board ship in 1770; and a finished watercolour prepared in England for publication as an engraving.**

**Spiny anteater, painted by a convict artist in Australia around 1790.**

them to crack the Iberian near-monopoly on transatlantic communications. Following a few undersupplied starts during the 16th century, when the expeditions had succumbed to the ravages of malnutrition and Indian attack, their efforts began in earnest in the 1600s, by which time smallpox and other diseases had already prepared the way. Spanish forays up the Mississippi River had exposed the indigenous population to deadly infection, and the new arrivals brought with them diseases of their own. A French outpost in modern Nova Scotia noted a disastrous epidemic in 1617 which scorched its way along the eastern seaboard – it was in all probability smallpox, but at any rate a preparatory bombardment of microbiological artillery that would prove of great assistance to the landing of the English Pilgrim Fathers in Massachusetts in 1620. Farther south the Indians were less depleted, and came close to overrunning the fledgling English colonies in Virginia in 1622. But there, too, the European pathogens were providing their terrible assistance.

Fighting between English and French colonists held back development at least as much as did the natural rigours of their new environment, but no such squabbles could restrain the advance of their allied species. East of the Mississippi there were no indigenous herbivores and, just as on the southern pampas, native grasses had not evolved to tolerate constant grazing. Once more, Europe's livestock munched and trod a way clear for Europe's plants, which in turn supported more European livestock and people. White clover – once honeybees had been introduced from Europe to pollinate it – and the so-called "Kentucky" bluegrass, among the most successful of the imports, did not so much accompany the westward-moving frontier as constitute it: they were themselves the frontier, crossing the Appalachian Mountains in the 18th century ahead of human pioneers. European weeds, the archopportunists of the plant world, thrived on the upheaval. Plantain, for example, earnt for itself the Indian nickname "Englishman's foot" by appearing everywhere the interlopers trod, and sometimes where they would not tread for a few years yet. For, like the bluegrass

and indeed the honeybee – "English flies" to the Indians – Old World weeds often swarmed ahead of the Old World people who inadvertently introduced them.

The colonists did not depend entirely on their own plants for sustenance: where the Amerindians possessed a useful food crop, Europeans were quick to learn how best to adopt it. If it was capable of growing in Europe, ships carried it back to the homeland. Thus maize – which was then known as "Indian corn" – had soon become a staple for the intruders as well as the natives, and rapidly became well established in Europe. Along with the tropical root called manioc, it was introduced into Africa, where it probably played a part in helping the population to increase despite the depredations of European slave raiders.

The cod fisheries on the Grand Banks off Newfoundland, which had first been reported by the English-sponsored navigator John Cabot in 1498, also proved a boon to the Europeans, and salt cod soon became a staple in the diet of parts of the Mediterranean region. Turkeys, too, made the eastward crossing, as did the potato, the most important American food plant of all.

The exchange of plants and animals was for the most part one-way traffic, however. Very few American weeds were able to establish themselves in Europe – though hundreds of their European counterparts hitchhiked their way to the New World. It was an imbalance that had contemporary botanists puzzled, but the reason was not hard to find: weeds thrive on upheaval, and it was in America, not Europe, that an ecology was in the process of being overturned.

On the microbiological level, the Americans may have passed syphilis to Europe: medical records are ambiguous, but the disease first appeared in virulent form shortly after the earliest voyages to the New World. But though it might have claimed many victims, syphilis did not possess the sheer population-destroying power of smallpox and the rest of the European contributions to American illbeing;

**Moorish idol, painted off the coral reefs of Tahiti in 1769.**

and venereal disease, whether of American origin or not, was to prove one of the final scourges of Amerindians when their cultures began to disintegrate. Like weeds, the germs were spread on wings of chaos.

Not all of the European imports to the continent contributed directly towards the downfall of the native populations. Horses, for example, breeding wild in Mexico, migrated northwards in enormous herds to settle in the Great Plains, where the small population of Indian buffalo hunters who lived there took to the art of horsemanship with enthusiasm. The animals completely transformed their lives. Buffalo hunting on foot had provided at best a marginal existence; mounted, the Indians prospered mightily. As the 19th century began there were at least 100,000 of them, following the movements of buffalo herds whose own numbers were probably between 30 and 40 million. By around the same date, however, the European colonies on the eastern seaboard had become a powerful, independent nation of almost five million people. Already they claimed the Mississippi River as their western boundary; and their citizens, covetously eyeing the vast territories that lay beyond, were increasingly well equipped to seize the new land for themselves.

The Europeans had not reached a position of such advantage without a struggle, but they had been aided beyond their own understanding by their nonhuman allies. It was as if the conquest of America, unplanned and opportunistic though it was, had been controlled by a brilliant general staff with a whole range of species working to its orders. First, a wave of microorganisms devastated the native people, then a second wave of larger life forms – pigs, cattle, corn – wrought no less dramatic changes upon the environment. It had taken the colonists' ancestors thousands of years to impose their familiar, human-centred ecology upon the Old World; it took only a few hundred to impose it on America.

For John Winthrop, who was the first governor of Massachusetts Bay Colony during the 17th century, the fact that the local population was "near all dead of smallpox" appeared as clear evidence that "the Lord hath cleared our title to what we possess." Governor Winthrop was not the only European who saw the hand of God at work in the American takeover, and indeed there were many who were convinced that divine support would ultimately afford the Europeans similar dominion over the rest of the world. But there were limits to the favour that the Almighty appeared to have granted Europe. The colonial accomplishment in the Americas may well have been resoundingly successful, but building equally successful colonies elsewhere than in a susceptible ecology whose climate was broadly similar to that of Europe's own proved a much tougher proposition.

Even in the Americas the Europeans did not have everything entirely their own way. The profits of the French fur traders, for example, depended not on imported European techniques but rather on the traders' adaptation of indigenous resources such as birch-bark canoes, moccasins and snowshoes. In the Spanish domains of Arizona and New Mexico, significant numbers of the local people not only managed to survive the impact of the European invasion but even retained their traditional ways of life largely intact. And although European plants and domestic animals had proved highly adaptable, wheat still could not be made to grow in rainforest conditions, nor would cattle thrive in a desert.

The colonists were not particularly interested in the vast tracts of desert, of course, but the seeming riches of the tropics made them another matter altogether: Columbus

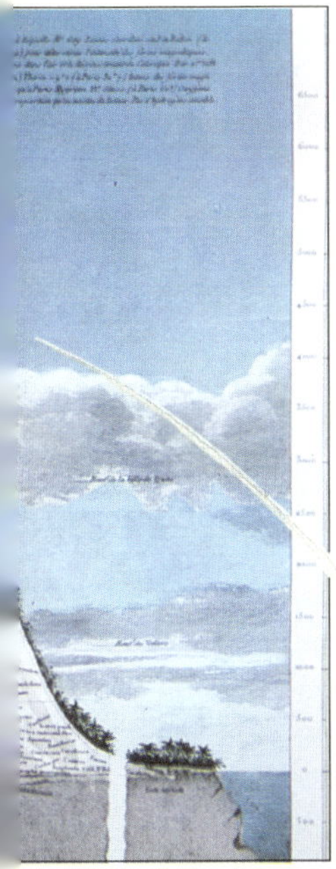

**The German naturalist Alexander von Humboldt is shown in this 1806 portrait *(far left)* seated over a plant press in South America, where he travelled for five years from 1799. Back in Europe, he published his findings in 30 volumes illustrated by engravings such as that on the left, which shows plant distribution in a volcanic region of Ecuador. In his major work *Kosmos* he sought to summarize the structure of the universe.**

himself had been greatly impressed by the verdant coastline of Venezuela, and explorers of other nationalities were quick to echo his enthusiasm. Attracted by the lush greenery and the apparent prospect it offered of an idyllic life style, many Europeans now made a determined effort. The Scots, who had been excluded by English law from the English successes in Massachusetts and Virginia, compensated by sending thousands of hopefuls to settle in Darien in the Isthmus of Panama, while for their part the French planted a handful of brave souls in Guiana. But the hot and humid tropics would generally prove themselves the graveyard of hopes and hopefuls alike.

In order to effect a successful wholesale transplantation of their ecology, Europeans needed to be able to make use of all of their advantages together to create an unbeatable combination: their weaponry, their diseases, their animal and vegetable allies, and the degree of evolutionary vulnerability of the particular environment they were seizing. Outside certain hospitable parts of the American continent, however, they found that they had no benefits but their technology and their rapacity at their disposal, and these proved to be insufficient.

The Portuguese, who had pioneered the very techniques of ocean sailing that opened up the planet, were the first to run up against the limitations of European power. Back in the late 15th and early 16th centuries, by dint of skilful seamanship, they had already found the passage to India that their merchants sought; their well-armed ships could usually count on crushing any opposition, and they were able to carve themselves a coastal enclave at Goa on the western coast of the subcontinent. But India had a dense population of pathogens – animals and successful crops as well as humans – all of which had coexisted for millenniums. And although the Portuguese managed to hang on to Goa, and indeed earnt massive profits from their Indian trade, their soldiers and merchants found the place dangerously unhealthy. There was no prospect of an ecological takeover on the transatlantic model.

The British and French colonization of India also fell a long way short of achieving complete subjugation. By 1800 the British had managed to build the foundations of a formidable political empire, but although in time they learnt enough prophylactic medicine to resist indigenous Indian diseases, they were never able to remodel their new territory as they had done America.

It was the same story throughout the rest of Asia. European sea power granted economic dominion, but did not bring with it the possibility of settlement in a biologically crowded – and resilient – environment. Although the European trading posts that studded Southeast Asia from the 17th century onwards often developed to become centres of political rule, they never became centres of European population.

**The parrot in the loving gaze of the Dutch housewife in this 17th-century painting *(above)* was probably brought from the East Indies, one of a number of foreign species that entered ordinary European households as pets following the voyages of discovery and the expansion of trading networks. For aristocrats, too, for whom the keeping of exotic animals was a long-established tradition of privilege, opportunities widened: the cheetah *(opposite page)* in a detail from a 1765 painting by George Stubbs was the first in England, a gift to George III from the Governor of Madras. Although in all countries the most widely available and popular pets continued to be dogs and cats – such as the kittens *(opposite)* in a detail from an 18th-century Chinese painting – the invasion of European domestic space by New World and Asian species brought a broader awareness of animal types and behaviour, and often a new curiosity and affection.**

Even the Philippines, seized by Spaniards sailing across the Pacific Ocean from Mexico around the middle of the 16th century, were never Europeanized.

Africa, too, was for a long time able to rebuff attempts at colonization – defended less by its people, though they had at least learnt how to smelt iron, than by its sicknesses. "A striking angel with a flaming sword of deadly fevers", as one explorer put it, guarded the continent against European humans and livestock alike. Though the Portuguese had set up bases on West Africa's offshore islands from the middle of the 15th century, and the French, the English and the Dutch all established trading forts during the 17th century, the interlopers had to endure appalling losses. Tropical Africa, especially West Africa, was a place where Europeans stayed for only the briefest of periods.

Only in the far south did Africa offer colonizing Europeans a climate that at least approximated to their ideal. The Dutch landed at Cape Town in 1652, and though the main purpose of the Cape settlement was as a stopping-off point for ships on the long trade routes to the East Indies, Dutch farmers and their livestock spread slowly northwards into open grassland they called the veld. Unlike the American pampas, though, the African veld was fully provided with grazing animals of its own – and human pastoralists, the Bantu, themselves part of a southward movement of African tribes. The fire power of their Dutch muskets guaranteed the settlers victory in most armed encounters, but the density of the local population was generally too low to make it susceptible to destruction by epidemic disease. The Dutch – just as would prove the case with their colonial successors, the English – could dominate, but not overwhelm; their progress was painfully slow by the standards of the European advance through the Americas. The Bantu survived, generally in the role of servants of the white intruders, and their numbers increased at a greater rate than did their masters'. South Africa never became an authentic transplant from the European stock.

Elsewhere in Africa most of the European visitors came only with the intention of making as quick a profit as possible and then departing. And in this aim they were often highly successful, for Africa possessed abundant supplies of one resource that on the other side of the Atlantic was in great and unceasing demand: its people. The massive depopulation of the New World in the early years of the 16th century left its Spanish exploiters without a workforce to labour in their plantations and silver mines. African slaves were the answer. From 1505, when the first wretched victims were dragged from their homelands and transported to Hispaniola, until the trade finally came to an end in the 19th century, at least 10 million Africans crossed the Atlantic in the slavers' holds. Many more of them died before they ever saw the New World, either in the prisoner convoys that brought them out from the interior of the continent, in the coastal barracks where they were held awaiting transport, or in the hideously overcrowded ships themselves.

Black slaves made large-scale sugar production in the West Indies and, later, Brazil, not just a possible but an immensely profitable venture. In the Caribbean islands

the Blacks would replace the original population entirely; in the mines and plantations of Mexico and South America they lived and bred alongside the surviving indigenes; in the southernmost of the English American colonies, by contrast, they remained a separate field-hand caste.

Apart from irrevocably altering the genetic mix of the Americas' future population, the slave trade had a curious and unintended consequence. As well as the involuntary passengers in the hold, some of the ships also brought stowaways in their water butts: mosquito larvae carrying with them the plasmodia responsible for yellow fever, malaria and a number of other insect-borne infections. Once ashore in the New World, the insects and their parasites soon sought out congenial environments in which they could multiply, and by the middle of the 17th century whole regions of South America – Amazonia, for example – which Spanish explorers had previously noted as disease-free became as lethal to Europeans as the West African rainforests. They would remain virtually inviolable, like most of Africa, until the advent of modern medical techniques in the late 19th and 20th centuries.

Writing of the voyage he undertook to the New World in the 1490s, Christopher Columbus enthusiastically described his discovery as "an earthly paradise". A century later the English poet John Donne playfully likened his mistress to that still unravished continent: "O my America! my new-found-land." And in the 17th century the philosopher John Locke remarked that "in the beginning, all was America." In fact neither America nor any other of the lands newly encountered and colonized

**Two views prepared by the English landscape gardener Humphrey Repton in 1790 illustrate his prospective patron's estate as it then was *(opposite)*, and Repton's proposal for its aesthetic improvement *(below)*. The principal changes suggested by Repton are the addition of a lake, to be created by damming a river; the conversion of cultivated fields to parkland studded by trees; and the construction of a mock-classical temple, peeping discreetly from the copse on the right. Undulating natural contours, previously abhorred, became fashionable in mid-18th-century England as a backdrop to the mansions of the rural gentry. Accordingly, landscape gardeners rejected the geometrical designs that remained popular in France and employed irregular lakes, terraces, sunken walls and artfully placed trees as a means of mediating between land used for private recreation and that put to productive use.**

by the European adventurers was virgin territory: they had all been home for thousands of years to human civilizations, and their inhabitants might well have responded, in the wry words of Miranda in Shakespeare's *The Tempest,* " 'Tis new to thee". But to excited and curious Europeans, the freshness of the new discoveries, and their apparent lack of recorded history, repeatedly evoked images of the biblical Garden of Eden. And just as Adam's first task in the original paradise had been to give names to the animals, so the Europeans set about naming, classifying and ordering the wealth of new life set before them.

The will to capture natural phenomena in the net of language, no less significant a form of taming the wilderness than agriculture or deforestation, was itself far from being new. Less legendary ancients than Adam, most notably the Greek philosopher Aristotle, had made determined attempts to discover order and pattern in the natural world; and Aristotle's systematic classifications of plants and animals, based on shared and not always obvious characteristics, had at least established that a dolphin was a very different sort of animal from a fish, and a bat from a bird. But the discovery of the Americas made clear that the ideas and even the knowledge of the ancients,

In a painting by English artist Joseph Wright, exhibited in 1766, a philosopher makes use of an orrery – a mechanical model of the solar system – to explain the theories about the operations of the universe proposed by astronomers and scientists such as Copernicus, Kepler and Isaac Newton. By turning a handle the philosopher could cause the planets to rotate on overlapping concentric bands around an object representing the sun, in this case a lamp wick burning in oil. Such public demonstrations made known to a wide lay audience – here including a woman and three children – scientific knowledge that, as the French writer François Voltaire had noted in the 1720s, was previously much talked about but rarely understood.

whose revival had helped lead to the Renaissance in Europe, were shockingly limited. The world was not only far larger than Aristotle had imagined, it was also far more complex, with a wider range of climate and crowded with plants and animals that rivalled anything to be found in the imaginary bestiaries of the Middle Ages. Among the peoples that explorers and colonizers encountered, there was an equally baffling diversity of race and culture. God's creation was even greater than had been supposed, his purpose perhaps more inscrutable.

Even without the added variety of the New World to take into account, the 16th-century classifiers were already having difficulty. The majority of them were medical herbalists – useful plants clearly enjoyed a high priority – and they often found it impossible to recognize plant species from others' descriptions, especially if they were northern Europeans trying to decipher a classical work which came from the different climate of the Mediterranean. In any case, an arrangement of plants according only to their medicinal properties was not likely to cast much revealing light on their natural relationships. There remained a crying need for a system of description and classification – a taxonomy – that would not only reflect reality but would also be universally understandable.

Observation had to precede any attempt at description, and for this it was clearly necessary to gather together representative specimens of as much of the new-found flora and fauna as possible. The first botanical garden in Europe was established at Padua in Italy in 1545, at least in part for the purpose of observing the new plants that were being brought back from abroad by the navigators, and other such gardens were soon flourishing elsewhere in Europe. Over the course of the following two centuries, English, Dutch, French and Spanish gardens were planted with exotic trees and other imported species, and these were complemented by the creation of zoos, glass-houses and museums for the study of animals, plants, seeds and rocks. George Stubbs, the celebrated English painter of horses, took time off from his equine routine to paint the first zebra and the first cheetah to be brought to England – from South Africa in 1762 and from India in 1764 respectively.

By the 17th century empirical observation and intellectual developments in many different fields had fermented together into what would later be called the scientific revolution, not least because of the invention of new instruments – among them the microscope, the telescope, the thermometer and the barometer – that had the ability to shed light for the first time on phenomena that were invisible to the naked eye. Physics was the greatest immediate beneficiary, but from among the herbals and bestiaries of the past the fledgling sciences of botany and zoology were fast emerging. Their practitioners, with an increasing amount of reliable information at their disposal and the advanced printing techniques that let them disperse it widely, were quickly learning how to classify their ever-larger number of specimens into groupings that shared multiple and complex characteristics.

In the 18th century the Swedish scientist Carl von Linné came up with a system of classification that sacrificed some of the "natural" categories of earlier rivals in favour of ease of use and consistency: genera and classes of life were established according to simple and often countable criteria. Linné also introduced the system of scientific nomenclature that has survived to the present day, virtually inventing a new dialect of Latin for the purpose, and it is in the Latin form of Linnaeus that his own name has been remembered. The Linnean system of classification allowed explorers easily to describe what they had seen in such a way that it would be equally easily understood by others, and a torrent of specimens, drawings and notes flowed back from all over the world to the universities, scientific societies and private studies of Europe. Meanwhile the French naturalist Georges Louis de Buffon was starting work on his encyclopedic *Histoire Naturelle*, of which he had managed to complete just 36 out of the proposed 50 volumes by the time of his death in 1788. But although his project was incomplete, Buffon had collated a wide range of previously unconnected data and used it as the basis to propose a revolutionary theory of geological history that would be further developed by Charles Darwin.

The overall effect of the new intellectual developments was to enshrine scientific knowledge and physical order as the gods of European civilization. Through understanding, Europeans also acquired a new confidence in their own authority over the natural world around them. The French philosopher René Descartes had defined man as a thinking being – animals, on the other hand, were nothing more than robots, incapable of either thinking or feeling pain. Scientists such as Galileo and Isaac Newton had asserted that nature obeyed mechanical laws and that, by understanding those laws, human beings had the power to predict events in the natural world. Moreover, to those who had been fully converted to the new doctrines, the application of science and human reason appeared to promise an attractive freedom from want, scarcity and the arbitrariness of natural calamities.

Within Europe itself, where the population began to increase rapidly from around the middle of the 17th century onwards – partly as a result of dietary improvements derived from the importation of New World plants such as maize and the potato – human ingenuity and muscle power were put to use to increase the production of food. Vast works of land reclamation and drainage were undertaken in northern Italy, western France, the Netherlands and the fens of eastern England. Deforestation in northern Russia cleared new land to be used for agriculture, and the untamed wilderness of Siberia gradually yielded to human settlement. Applied to stockbreeding in western Europe, scientific knowledge aided the artificial development of new breeds of cattle, sheep and pigs that would yield more fat and meat than their forebears had done. In Britain the invention of new agricultural machinery – such as Jethro Tull's mechanical sower, which planted seeds in neat rows – enabled the labour of planting and harvesting crops to be carried out by fewer men in a much shorter time than before. Such technological developments were harbingers of the industrial revolution which in the 19th century gave human beings the means to let them exploit their natural resources with greater efficiency than even the most anthropocentric of the early naturalists had ever imagined possible.

Outside Europe, too, science was altering human attitudes to nature and providing the impetus for further development. The explorers who were involved in the second great age of European discovery during the 18th century, unlike their 15th-century equivalents, were generally naval officers who sailed on the orders of their government, in well-equipped ships which carried scientific observers and sophisticated instruments for measuring and recording what they found. When the Royal Navy's Captain James Cook, the most painstaking navigator of his time, made landfall on Australia's habitable eastern coast in 1770, he was accompanied by the eminent botanist Joseph Banks, and their primary concern was to record and communicate hard scientific data about their discovery. Cook's own sympathetic description of Australia's Aborigines – "They may appear to some to be the most wretched people on earth," he wrote, "but in reality they are far happier than we Europeans" – reflected a new attitude towards the people of other cultures, which had its roots more in curiosity and interest than in the unrestrained will to dominate previously evident.

In the event, scientific objectivity was not enough to blunt the ecological impact of the Europeans on a continent that had been separated from the main planetary land-mass for even longer than America. The Aborigines of Australia, hunter-gatherers who possessed no weapon more powerful than a stone-tipped spear, no domestic animals other than hunting dogs and no epidemic diseases whatsoever, were even less capable of defending themselves and their environment than the Amerindians had been. Following the establishment of a British penal colony at Sydney Cove in 1788, the colonization of Australia proved to be a recapitulation, on a compressed timescale, of the American experience three centuries before. The colony's governor went out of his way to avoid harming the original inhabitants, but his conscientiousness was to no avail. By 1789 Aborigines who were in contact with the newcomers were dying of something akin to smallpox, even though the long ocean voyage of the convict settlers should have served as an effective quarantine. Passed from band to

**The knowledge won by scientific advances in the 17th century increased the human capacity to modify and – as the advocates of progress insisted – improve the given shape of the natural world. Challenged by the demands of populations that in Europe began to rise steeply from around 1650, agriculturalists in Britain and the Netherlands especially put this power busily to use. Smallholdings were amalgamated and enclosed to create compact farms that were economical to run; estate owners experimented with new crops and systems of crop rotation, and invested in mechanical sowers and factory-made ploughs; the size of livestock was increased by scientific methods of breeding; and new land was claimed for agriculture by extensive drainage. The panorama on the right and extending overleaf shows an estate in Gloucestershire in around 1730: behind the teams of labourers gathering hay into stooks and advancing in line with scythes, more distant fields are patterned by the regular ridges and furrows left by ploughing. In England alone, 2.5 million hectares were enclosed and worked in this manner between 1700 and 1845.**

# IMPROVING ON NATURE

band, the disease swept inland along the hunting routes, killing perhaps a third of Australia's indigenes before the English had even moved out of sight of the sea.

As in America, European livestock continued the conquest. The eight beasts that had been bought by the 1788 British fleet at a Cape Town stopover escaped into the hinterland. By 1804 the strays had bred into a herd of 5,000 wild cattle, at once a resource for the growing colony and a biological bulldozer that opened niches for European plants and people as it spread and reproduced. The Aborigines, whose hunter-gatherer culture demanded vast spaces, would be squeezed towards extinction. Within a few human generations, the cattle would be joined by millions of sheep, their wool destined for Britain's newly mechanized textile industry. Australia would be a European-style state with an English-speaking population millions strong.

By 1800 the expanding European ecology was nearing its natural limits. Of all the temperate lands whose inhabitants might conveniently be dispossessed, only the islands of New Zealand, which had been mapped by Captain Cook but were devoid

of White settlement until 1814, remained. The great, if rather one-sided, exchange of plants and animals which had started in the time of Columbus over 300 years before was almost complete: even before the arrival of the alien settlers, New Zealand's native Maoris were noting with consternation the strange new weeds that were already springing up on their islands. Potatoes were fast becoming a staple in Europe itself; maize and manioc were firmly established in Africa. The European hegemony over the seas had arranged the transfer of other plants, too: breadfruits, for example, from the islands of Tahiti and Hawaii, whose indigenous populations were now suffering their first experience of epidemic disease, had been naturalized in the West Indies as a cheap nourishment for plantation slaves.

But even while the ecological explosion that had been detonated by the impact of Europe on the wider world was beginning to subside, the narrative of human adventure was still a long way from reaching any satisfactory conclusion. The domination of a previously very local, Europe-centred civilization over a large part of the globe's terra firma was now ensured, and had been achieved in what was a remarkably short

**In a painting dated 1810 *(above)*, the English stockbreeder Robert Bakewell hires out rams to farmers seeking to improve their herds. By methodical selection and culling, Bakewell developed early maturing sheep with a high yield of quality meat.**

period for the task, but the new overlords' comprehension of the land to which they had laid claim still lagged a long way behind their actual expansion.

Emblematic of this failure in understanding was the experience of the expedition led by Meriwether Lewis and William Clark which set out in May 1804 from St. Louis on the Mississippi River, intending to reach the western edge of the American continent. Comparable in ambition to Columbus' transatlantic venture three centuries before, the Lewis and Clark expedition, sponsored by President Thomas Jefferson, left base armed with the most up-to-date equipment that was available: scientific instruments, vaccines, friction matches, textbooks on botany and mineralogy, an air gun, a collapsible boat. They were imbued, too, with the most enlightened ideas of the time, instructed by the President to treat the Indians well and to make detailed records of climate and of flora and fauna "with great pains and accuracy". But despite all their apparent advantages, they were unprepared for the landscape they traversed, which seemed to mock their puny baggage. Assailed by storms, cactuses, mosquitoes, rattlesnakes, cougars and grizzly bears, compelled in desperation to survive off melted snow and the flesh of their own horses, Lewis was driven to report that "to make any further experiments in our present situation seemed to me madness." The expedition did eventually reach its goal of the Pacific coast, but for his remaining years Lewis was recurrently subject to "sensible depressions of the mind". He died in 1809 of gunshot wounds, probably self-inflicted.

But if the European expansion had proceeded at a pace that outreached European understanding, it was now too late to slow it down. Industrial development and the growing population of Europe served instead to accelerate the momentum, and in the ensuing decades of the 19th century the westward route of the Lewis and Clark expedition was to be followed by wagon trains bearing thousands of hopeful settlers. Overcoming their preconception that only that land where trees grew in abundance was properly fertile and suitable for agriculture – a belief that had already led to massive deforestation – and greatly aided by the steel plough, barbed wire, repeating rifles and advancing railroads, newcomers from Europe inexorably turned the Indian hunting grounds of the Great Plains into a profitable granary, and marched west across the Rockies in quest of Californian gold. They were just one contingent among some 50 million emigrants who crossed the oceans from Europe during the century, the majority of them to the United States but many also to the pampas regions of South America and to Australasia. These areas had a total population of little more than five million as the century began, far less than the 12 to 15 million of the pre-Columban Americas alone; natural increase combined with immigration had made that number close to 200 million in 1910.

By then, something approaching 90 per cent of the population of the United States was of European descent, with Africans accounting for most of the remainder and the descendants of the original Amerindians making up only a tiny fraction. But the numbers alone were only one indication of the great changes that had taken place. Even more significant, the uses to which the land and its resources were being put were now determined by cultural attitudes which were completely different from those of its original inhabitants. "We did not think of the great open plains, the beautiful rolling hills and winding streams with tangled growth as wild," declared a Sioux Indian chief. "Only to the White man was nature a wilderness. To us it was tame. Earth was bountiful and we were surrounded with the blessings of the Great Mystery." Now that great mystery was being eroded.

# THE ODYSSEYS OF PLANTS

Plants provide the necessities of food, shelter, fuel, clothing and medicine; they are also used in the manufacture of such diverse things as furniture, paper, pigments, cosmetics and photographic film. Among the most precious resources of any society, they have always excited the envy or greed of those who have not possessed them, and their dissemination from one part of the globe to another has long been speeded by purely political no less than by natural currents.

All the plants shown on the following pages now grow in countries to which they were first taken by human agents, often as trade goods or following in the wake of military conquests. Their movement around the world has altered natural landscapes and revolutionized national economies. The coffee plant, for example, which is shown *(right)* in a 17th-century French engraving, originally grew wild in Ethiopia, from where it was introduced by traders to Arabia during the 15th century. Coffee beans were imported into Europe along with a wide range of Oriental spices, and the popularity of the drink prompted the European colonial powers in the 17th and 18th centuries to cultivate the plant in whichever of their overseas possessions proved suitable. A single bush was first carried to Martinique in 1723; from the Caribbean islands, coffee plantations later spread to coastal areas of South America and Brazil, which is now the largest producer of coffee in the world.

This and the other plants here are a mere sample of those spread by humans from their original habitat. Food plants alone include the tomato, the banana and the chilli pepper. And no doubt there are more that still await dissemination: studies of tropical rainforest plants, for example, show that many contain chemical compounds new to Western science that have potential for use in medicine and industry.

A 16th-century engraving *(right)* shows a sugar mill in Sicily: juice from sugar cane is pressed out, boiled and poured into conical moulds to set as loaves. Sugar cane, shown below in an illustration from a 1615 treatise, originated in India and was introduced to Europe after the Muslim conquests of the seventh century. Columbus carried seedlings from the Canaries to Hispaniola in 1493; within a few decades most of Europe's sugar originated from the West Indies.

A Peruvian pot *(far left)* dating from around the 10th century is shaped in the form of two potatoes, which grew wild in South America as early as 6000 BC. Carried to Spain during the 1530s by the conquistadors, the potato at first fetched high prices as a cure for impotence, but became a common European food by the late 18th century. The failure of potato crops in Ireland in 1845 caused widespread famine, illustrated by the 19th-century engraving *(left)* of women and children searching for potatoes that had escaped the blight; starvation and emigration reduced Ireland's population by nearly a quarter in the following six years.

Dating from around the 14th century, a life-size cob of maize with silver grains and gilded husks *(far right)* testifies to the plant's importance to the Inca of Peru. The dietary mainstay of all early Central and South American peoples, maize was first cultivated in around 5000 BC. The plant was taken from the Caribbean to Spain by Columbus in 1496; a century later it was planted throughout Europe and in parts of the Middle East. On the right a detail from an 18th-century Venetian painting shows a woman turning out a pot of polenta, made from yellow maize flour, which became a staple food of the poor in northern Italy.

A picture of the peanut plant *(left)* in a book printed in the Philippines during the 19th century shows its unique growth pattern: once fertilized, the flowers burrow underground, where the seeds then ripen in brittle shells. Native to tropical South America, the plant was taken to Europe by Spanish explorers and then to Africa and Asia. Brought to North America as food for slaves, it became popular in the Civil War, when Union soldiers – as shown in the painting by Winslow Homer *(below)* – roasted the nuts on campfires. Except in the United States, the plant is grown mainly for its oil.

An Egyptian tomb painting of 1200 BC *(below)* shows the harvesting of flax, whose fibre was used to make linen for clothes and to wrap the bodies of the dead. Flax was also grown in prehistoric Europe, but not until the Romans introduced linen to France and Belgium was it used to make clothing on a large scale. On the right, an illustration from a 15th-century Italian edition of the work of the Roman natural historian Pliny the Elder shows women separating flax fibres from the woody stems. Linen remained Europe's most common clothing material until the advent of cheap cotton in the 19th century.

In a game recorded by a French artist in 1591, Central American Indians bounce rubber balls made from the milky sap of a native tree. After noting local products derived from another species of the tree native to the Amazon Basin, Europeans first used rubber for waterproofing fabric and for tyres in the early 1800s. Plants grown at Kew Gardens in London from Brazilian seeds were sent in the 1870s to Ceylon and Malaya, where large plantations were grown. On the opposite page, Henry Ridley, director of Singapore's botanic gardens from 1888 to 1911, demonstrates the rubber-tapping method he devised: a herringbone pattern of incisions that ensures an easily collectable flow of latex while allowing the bark to renew itself smoothly.

Fabric made from the cotton plant – depicted in a 19th-century Indian drawing *(right)* – has been found in tombs in Peru and India built before 2000 BC. The plant later became widely dispersed: it grew around the Mediterranean in Roman times, reached China in AD 600, and was introduced to Africa by the Arabs in the eighth century. Cultivation in North America began early in the 17th century; the labour of African slaves, shown in the 19th-century engraving above, and the invention in 1793 of the cotton gin, which sorted the seeds from the fibre, made the U.S. the world's largest producer.

In an early 19th-century painting *(above)*, Dutch merchants sample tea in a Chinese warehouse. In legend, tea was drunk in China as early as 2700 BC. By the eighth century, tea plants grew in Japan and other countries of Asia. Imported from Java to Europe by the Dutch in 1610, tea rapidly became a popular drink in Holland, England and the American colonies. The British, who found wild tea in India, began cultivating the plant there in 1834; by 1900, 764 plantations covered the hills of Assam *(right)*.

The tobacco plant derives its name from an Americar Indian word for a leaf cylinder for smoking – shown in the 16th-century Dutch engraving *(far left)*, the first published picture of the plant – or a tube fo inhaling the smoke. In the 1550s the plant, native to tropical regions of the Americas, was brought to Europe, where it was first prized as a cure-all for everything from snake-bites to ulcers. Smoking was popularized in England by Sir Walter Raleigh; most Europeans smoked pipes, as shown in the illustration *(left)* from a 17th-century Dutcl pamphlet, but the Spanish preferred cigars. Grown by European settlers in North America from 1612, tobacco became the main commodity traded by the colonists for European goods. Tobacco is now grown in about 80 different countries.

In an engraving from the 19th century, bark is stripped from cinchona trees in the forests of the South American Andes, the tree's native habitat. The protection against malaria afforded by quinine, a drug derived from the bark, was first made known to Europeans by Jesuit missionaries returning from South America in the 17th century. The Dutch and the British later grew their own supplies in Indonesia, Africa and India, where quinine facilitated their conquests and was sold in packets *(inset)* to colonial officials.

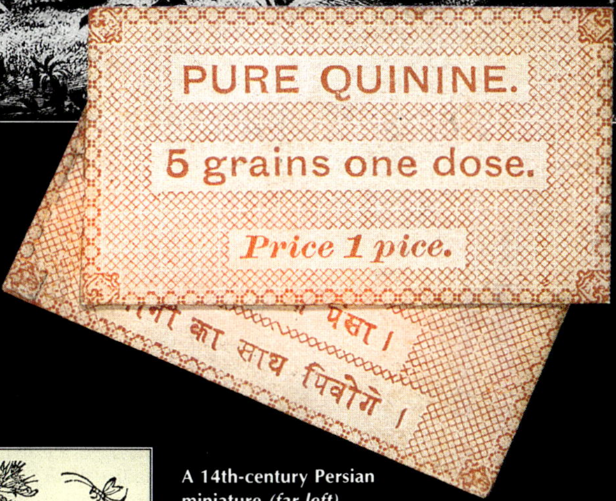

A 14th-century Persian miniature *(far left)* shows a lady enjoying the intoxicant effects of cannabis. The plant, shown in a 17th-century French drawing *(left)*, originated in central Asia and was grown in China by 2700 BC. It was cultivated throughout Europe in the Middle Ages, in Chile from the early 16th century, and in British colonies in North America a century later. Also called Indian hemp, it is grown for its tough fibres – used to make cloth and rope – as well as for the mildly hallucinogenic drug made from its resin. The plant was once used as a sedative and analgesic, but has no known use in modern medicine.

# THE GLOBAL CHALLENGE

6

One day during the 1930s a motorist was driving on a country road across the Great Plains of the United States. Seeing a ten-gallon hat resting on a nearby sand dune, he stopped his car and got out. Under the hat, so the story goes, he was surprised to discover a man – or at least the head of a man. "Looks as though you're in trouble," said the traveller. "Let me give you a lift into town." "I'll get there myself, thanks," replied the head from its pile of sand. "I'm on a horse."

The situation behind this story was in fact the worst ecological disaster that the United States had experienced, a disaster of such magnitude that rueful humour was the only way many Americans could cope with it. Overcultivation was part of the cause. Posters during the First World War had urged: "Plant more wheat – wheat will win the war." The farmers of the southern Great Plains had done so, and in the postwar years they continued to plough up the fragile grasslands of Kansas, Oklahoma and northern Texas. Then, in the early 1930s, the rains stopped and the land grew bone dry. With nothing to anchor it, the precious topsoil swirled across the plains, bringing agriculture to a standstill and sending thousands of impoverished farmers on a pitiful exodus to California.

The blame for what had happened was difficult to apportion. The farmers claimed it was the fault of the politicians, and the politicians pointed in turn to the American public, whose appetite for cheap bread encouraged the sort of agriculture that could rapidly transform a sea of living grass into a sterile desert. But in a sense the responsibility was even more widespread, for the tragedy of the Dust Bowl – as this region was to become known – was just one of the many painful lessons that humanity throughout the world was being forced to learn.

Environmental crises were by no means a new phenomenon: erosion, pollution and the extinction of entire species had for centuries tarnished the achievements of civilizations. During the 19th century, however, the scale of humanity's impact on the planet increased to a scale where it began to assume menacing dimensions. Industrialization, technological developments and a massive growth in the world population confronted peoples everywhere with sobering evidence of environmental deterioration, from lifeless waters and barren hillsides to the asphyxiating air of smoke-choked cities. Voices were raised in protest, but it was difficult to argue against the powerful assumptions of progress and growth that governed contemporary economic and political thought, and industrial civilization hurtled into the 20th century with a reckless momentum that threatened ever-increasing catastrophes. Margins for error were progressively reduced. Now a simple accident had the potential to turn an ocean black with oil or fill the air with a deadly cloud of poisonous vapour. Nuclear weapons promised destruction on an inconceivable scale. A disturbing new lexicon of ecological concepts entered the vocabularies of ordinary men

**A helicopter flies above the devastated flanks of Mount St. Helens in Washington, U.S., spraying fertilizer on grass seeded to limit erosion after a massive volcanic eruption on May 18, 1980. The eruption tore more than 400 metres off the summit of the 2,950-metre-high volcano, destroyed tree cover over an area of 600 square kilometres, and claimed the lives of 60 people. "What we have here," declared a biologist surveying the grey wasteland, "is almost a Genesis situation." As such, it demonstrated nature's own regenerative powers. The deep roots of fireweed plants sent up shoots that broke the ash-encrusted surface; browsing animals and insects feeding on rotting debris gained a foothold, and rains restored the soil's fertility.**

**An 1837 lithograph shows labourers at work at the mouth of a tunnel on the railway line between London and Birmingham, which was opened in the following year. Unlike roads, railways had to be built as level and as straight as possible: experiments made in 1833 showed that a train ascending a gradient as shallow as 1 in 300 needed twice the tractive power required to haul the same load over level ground. Therefore hills were levelled or pierced by tunnels, and valleys filled in or spanned by soaring viaducts. Some 20,000 men laboured for five years to construct the 200-kilometre London to Birmingham line, a task that was compared by a contemporary to the building of the pyramids in ancient Egypt.**

and women: ozone depletion, desertification, nuclear fallout, global warming, acid rain. And as the century drew to a close, the earth's ability to sustain human life under any circumstances — a premise which past civilizations had never doubted — was increasingly seen as a dangerous illusion.

In around 1650 there began a sustained and rapid surge in the size of the planet's human population, the consequences of which would bring an alteration in the face of the globe itself. At that time the earth was home to around 500 million people — the number had doubled roughly every 1,500 years since 8000 BC. But over only the next two hundred years the population doubled again, reaching one billion by 1850. Most of this growth was centred in Europe, whose population rose from 140 million to 266 million between 1750 and 1850. The inhabitants of Britain and Ireland alone tripled in number during this period, in spite of successive waves of emigrants who fled to the more open spaces of America.

This rapid population increase was initiated by improvements in agriculture and by a growing spirit of free enterprise. No longer bound to the land as peasant farmers, Europe's rural masses flocked to the cities, ambitious for work and wealth. Many were to die in wretched poverty, but the sustaining glimmer of urban riches seemed always within their reach, and its flame was further fanned by tales of wealth from beyond the seas. Indeed, there was soon evidence of the New World's bounty, not just in the gold and silver — or codfish and pelts — that flowed east across the Atlantic,

**In the foreground of *Mountain Landscape with Rainbow*, painted in *c.*1810 by the German artist Caspar David Friedrich, a lone spectator stands in rapt contemplation of the majesty of nature. Artists and writers of the Romantic movement gave definitive expression to the new attitudes towards the natural world that had been evolving since the voyages of exploration and the scientific discoveries of Copernicus, Galileo and Newton in preceding centuries: the immensities of nature itself now rivalled God as the proper object of human awe. Like high-class tourist brochures, canvases such as that shown here prompted the leisured classes to visit the Alps, the English Lake District or the Scottish Highlands, their journeys facilitated by Europe's expanding network of railways.**

but in the new food crops, particularly potatoes and maize, which in turn helped to support even greater numbers of people.

As Europe's population expanded, more land was required for growing crops. The threat to the forests of England was particularly pressing. Substantial stands of timber had to be preserved for the maintenance of the navy, and wood was also the prime fuel for industry and domestic heat. The only alternative was coal, an organic mineral rich in carbon, and a resource in which England was extremely wealthy. The scale of mining had been steadily increasing since medieval times, and by the end of the 17th century most of English industry, from brewing to brick making, was fired by coal; only iron smelting remained dependent upon charcoal – which was derived from the burning of wood – and this last bastion of traditional industry toppled in 1709 when Abraham Darby developed a technique for using coke to fire his furnaces in Shropshire. Although the full impact of these developments did not become apparent for more than a hundred years, the Industrial Revolution was under way. It would alter for ever humanity's relationship with the natural world.

The mining and transport of coal tested the ingenuity of English inventors and engineers. Shallow opencast pits were quickly exhausted, and because deep mines tended to fill up with water, the need arose for an efficient pump. The steam-engine was first developed for this purpose, and soon proved to have a wide range of other applications. Coal had to be carried in bulk from the mines to centres of industry, and the use of horses to pull both wagons and canal barges was inefficient and wasteful.

**A pair of flightless great auks are captured in characteristic poses in this hand-coloured engraving from *Birds of America*, a pictorial survey published between 1827 and 1838 in which the artist James Audubon sought to record every living bird species in North America. Great auks, which for centuries had been protected by the isolation of their native habitat along the coast of Newfoundland, were killed in increasing numbers after the arrival in the 16th century of European fishing fleets, which provisioned their return voyage across the Atlantic with the birds' meat and eggs. A little more than a decade after Audubon sketched these and other birds on a trip to Labrador in 1833, the great auk was extinct.**

"It has been said that in Great Britain there are above a million horses engaged in various ways in the transport of passengers and goods," reported a select committee of the House of Commons in 1833, "and that to support each horse requires as much land as would upon an average support eight men." The solution lay in steam-powered trains, first introduced in 1825. By the middle of the 19th century Britain was bound in a growing web of iron rails.

The revolution took on a steady momentum of its own, one invention sparking off another like a series of explosions on a string of firecrackers. In the 1850s the English engineer Henry Bessemer devised an economical means of converting pig iron into steel. By 1859 the United States had begun to tap its underground reservoirs of oil, the fuel that was destined to transform once again the development of transport and industry. Innovations in the electrical and chemical industries opened yet more doors in the last quarter of the century.

The harm caused to the natural world was soon apparent. Industrial waste and untreated sewage poisoned rivers and canals. Factory chimneys spewed out sulphur dioxide, produced by the burning of coal, which defoliated the surrounding countryside. Cheap housing for the armies of new factory workers sprawled across former farmland. Villages that had scarcely changed in the previous century became lost overnight in a labyrinth of narrow streets and sooty brick. Farther afield, valleys were flooded to serve as reservoirs for the new conurbations, while the spreading railways steadily opened up new regions for development.

Although France was rapidly following England's lead in methods of production, the historian Alexis de Tocqueville was quite unprepared for the sheer ugliness of Britain's industrial heartland. "Thirty or forty factories rise on the top of the hills," he wrote of Manchester in his *Journey to England and Ireland*, published in 1835. "The wretched dwellings of the poor are scattered haphazard across them. Round them stretches land uncultivated but without the charm of rustic nature, and still without the amenities of a town. The fetid, muddy waters, stained with a thousand colours by the factories they pass, wander slowly round this refuge of poverty." Not everyone shared de Tocqueville's dismay. To many 19th-century Europeans and Americans the Industrial Revolution marked an important advance in human civilization. There was something inherently noble in the power that human beings could now harness for their own enrichment. "Examine the endless varieties of machinery which man has created," wrote one English enthusiast in 1831. "Mark how all the complicated movements cooperate, in beautiful concert, to produce the desired result. . . . We do entertain an unfaltering belief in the permanent and continued improvement of the human race, and we consider no small part of it, whether in relation to the body or the mind, as the result of mechanical invention."

Such sentiments were to have global consequences. The United States, France and Germany were early converts to the doctrines of industrialization, followed by Russia and Japan later in the century. And because Europe could not supply either its voracious new industries or its expanding population solely from its own resources, goods from abroad became essential buttresses of the Old World's economy: furs from Canada, cotton from the American south and India, beef from the Argentine, wool from Australia, hides, silk, jute and rubber. In many cases these goods answered a genuine need: because sheep consumed large areas of valuable farmland, for example, England's long-established wool industry was forced to contract, and by the middle of the 19th century British working men and women were clothed almost entirely in imported cotton cloth. But with rising standards of living, need accumulated upon need, playing havoc with both the economies and the environment of the non-European world. In addition, the European capitalists were committed to opening up new markets for themselves abroad – and the same ethic that permitted the exploitation of both labour and land in Europe also encouraged them to look upon the undeveloped world as little more than a giant factory, fit only for the production of food and the supply of resources.

**This sperm whale's tooth from the Pacific Ocean was carved with a picture of a whale being harpooned in the mid-19th century, when America's whaling fleet numbered more than 700 vessels. Such artefacts were produced by crewmen during their long periods of idleness as they sailed to the whaling grounds or waited for their prey to blow. Unlike the Inuit of eastern and western North America, who exploited every part of a whale's carcass for food, fuel, clothing and tools, American hunters were interested only in the whale's blubber, a major source of oil for lighting and heating before the discovery of petroleum oil in 1859.**

The traders and entrepreneurs who travelled outside Europe were not immune to the wonders of their new surroundings. "The larger portion of this desolate region is covered with lofty and luxuriant green forests in every direction," reported two British engineers who explored India's mountainous southwestern region in 1817. "The whole scene is truly sublime." But the lure of potential profit took precedence over beauty, and by the end of the century extensive tea and coffee plantations covered much of the mountainsides. In Assam, lying at the opposite corner of the subcontinent, 764 plantations were exporting 66 million kilograms of tea annually by 1900, while a great army of 400,000 low-caste workers, living in virtual slavery, ensured that the tea caddies of Britain were never empty.

The transformation of agriculture in the undeveloped world was not always the result of European occupation. Even before colonial rule came to West Africa, for instance, native farmers had acknowledged the commercial advantage of growing cash crops for Europe. A swift steamship service, begun in the 1860s, carried their produce northwards: peanuts, palm oil and cocoa. And the practice of growing crops for sale to the industrialized world, to the neglect of essential food crops at home, lasted through colonial days and into independence, sometimes at great cost to the environment. Silt from deforested hillsides clogged lowland rivers, watercourses flooded with destructive regularity, while foraging cattle gave new tree cover no chance to

establish itself. "All progress in capitalistic agriculture", Karl Marx grimly observed, "is a progress in the art, not just of robbing the labourer, but of robbing the soil."

Other agents besides humans had a devastating effect. Wherever a ship sailed into an undiscovered bay, or a convoy of wagons lumbered over a trackless plain, there came a living army of which human beings were only a minority. Rabbits were introduced to Australia in 1859; within a century their numbers reached an estimated 500 million. Pigs, dogs, cats, goats and the adaptable honeybee likewise all thrived in the new worlds to which they were transported. And wherever European men and women went, that most faithful of all domestic animals – the rat – was at their side.

Vermin and overgrazing by European livestock created conspicuous environmental problems in newly explored countries. The native plants were forced into retreat,

**Two white overseers lounge by the side of a track through a banana plantation in Angola in around 1890. To alleviate the financial burden of administering their African colonies in the late 19th century, European powers encouraged the growing of cash crops and laid down a transport network to speed their export, thereby subjecting the land to the same spirit of capitalist enterprise that had already left its mark in Europe. The railway *(opposite page)*, flanked by some of the 32,000 Indian workers who were imported to construct it between 1896 and 1901, was laid over nearly 1,000 kilometres of difficult terrain between Mombasa on the coast of Kenya and the northeastern shore of Lake Victoria in Uganda.**

often to the point of extinction. Thirty-three plants unique to the island of St. Helena were identified in 1810; since then, goats have accounted for the disappearance of all but 11 of them. Native animals were also threatened. Rats and pigs ate the eggs and young of ground-nesting birds. The dodo, native to the islands of Mauritius and Réunion, became extinct by the late 17th century. The takahe, a similar flightless bird of New Zealand, was unable to compete for grazing with the European red deer and only narrowly avoided suffering the same fate. In Australia, an expedition in the Murray-Darling region northwest of Melbourne discovered 31 species of mammals between 1856 and 1857; only nine of these survive.

A quieter but no less remarkable revolution was occurring in the vegetation of the colonized lands. Some plants arrived as official immigrants: more than 200 species were deliberately introduced to Australia by 1803. Hundreds more arrived as stowaways, their seeds mixed in with others or clinging on to the coats of animals. Their spectacular success literally changed the face of much of the world. Only a quarter of the plants growing on the Argentine pampas today are native species. Of the principal weeds in North America, more than half are of Old World origin. Only in the tropics did native species hold out successfully against the aggressive immigrants.

The most unwholesome allies of the European colonists were their own pathogens. Smallpox was less widespread in the 19th than in earlier centuries, but the invaders still maintained a formidable arsenal of germs. "Wherever the European had trod, death seems to pursue the aboriginal," observed the English naturalist Charles Darwin in 1839. New Zealand's Maoris, a determined and populous warrior people, fell in their thousands to tuberculosis and measles, while European venereal disease kept their birth rates artificially low. By the end of the century their numbers were down to around 42,000, from the nearly 200,000 who had occupied the islands on Captain Cook's arrival in 1769. A melancholy proverb that was circulating among the Maoris

**Goats and cattle forage along the shifting border between desert and pasture in Niger, West Africa. On the right *(inset)*, a wood vendor heads for a local market in Senegal, where timber is the only fuel for nine out of 10 rural households. Overgrazing and tree cutting, which can render fertile land barren, contribute to the encroachment of desert on the arid but potentially productive territories that comprise more than a third of the world's land surface.**

during the mid-19th century summed up the great problem that was facing aboriginal people throughout the temperate world: "As the white man's rat has driven away the native rat, so the European fly drives away our own, and the clover kills our fern, so will the Maoris die before the white man himself."

"I wish to speak a word for nature, for absolute freedom and wildness," announced Henry David Thoreau to a gathering in his native Concord in Massachusetts in 1851. It was a subject with which the young man was well acquainted. A keen naturalist

and woodsman, he had emerged four years earlier from a solitary life in a log cabin on the shores of Walden Pond, an experience he was to describe as "living deep and sucking all the marrow out of life". Thoreau's self-appointed mission was to persuade his fellow Americans that all of nature was a source of inspiration without which man's whole existence was superficial and empty. "In wilderness is the preservation of the world," was the rousing climax of his speech in Concord that day.

Thoreau's tone of impassioned urgency was an appropriate one, for it was in his own nation of the United States of America that the wilderness was fast succumbing to the effects of industry and material progress on a greater scale than anywhere else. Alexis de Tocqueville, during a visit he made to the United States in 1831, was astonished by the callousness and brutality with which the Americans treated their natural surroundings. "They are insensible to the wonders of inanimate nature," he complained, "and they may be said not to perceive the mighty forests that surround them till they fall beneath the hatchet." During the 1840s there were around 800,000 hectares of redwoods, the earth's tallest living things, growing in California; by the time their systematic felling was finally halted in the 20th century, 96 per cent of the mature trees had vanished.

Trees were for felling, animals and birds for shooting – and the rifles that were used to slaughter boars by the thousand on the great hunting estates of northern Europe had an even more dramatic effect in America. In the early 19th century no American bird was more numerous than the passenger pigeon. Hundreds of millions of these creatures darkened the skies in enormous flocks that might take two or three days to pass overhead. Early frontiersmen and farmers shot them for the pot, but with a hungry new market in the eastern cities, the culling of pigeons became a massacre. The birds were caught in nets, drugged with alcohol-soaked corn, asphyxiated with smoke or simply blasted out of the sky. Their bodies were loaded in their thousands into freight cars and hauled back to the markets of Philadelphia or New York. In 1878 one hunter in Michigan boasted an annual bag of three million birds. Within a decade their numbers had declined below a sustainable level. The last of the species died in captivity in a zoo in Cincinnati in 1914.

A similar story was unfolding on the Great Plains. An estimated 30 to 40 million American bison – or buffaloes, as they were popularly called – ranged the short grasslands of America's western territory at the beginning of the 19th century. Even equipped with rifles and mounted on horseback, the plains Indians who hunted them had made no significant impression on their numbers. Nor did they want to, for their way of life depended upon the buffalo's survival. To the White hunter, however, the buffalo was a crop to be harvested. The slaughter began in earnest in the 1860s and reached its height between 1870 and 1875, when an estimated 2.5 million buffaloes were killed each year. Their hides were baled up and sent off to tanneries; their bones

Three watercolours from a 1608 treatise by an Italian landowner, Guiseppe Paulini, show the effects of deforestation on a hill near Venice. After felling the trees, farmers fired the stumps to make fields; topsoil, no longer anchored by the tree roots, was washed away by rain, leaving the hillside barren.

This view of the Grand Canyon, painted in 1872, helped persuade the U.S. government to create Yellowstone National Park.

In the 19th century, human impact on the natural world was suddenly speeded up: vistas of untrammelled wilderness and of plains and valleys which had for centuries been marked only by the farmer's plough now increasingly bore the vivid scars of industrial development. For many this acceleration prompted a late concern for what was being lost, and a commitment to preserve what was left.

For some the value of untamed nature was perceived as spiritual: without wilderness, they claimed, the human soul would be diminished. For others its worth lay in the timber, soil and minerals that supplied the needs of civilization. The lesson that these resources were finite had been learnt and relearnt by every urban society since the Sumerians in Mesopotamia in 3000 BC; and the ruinous ways of humans themselves were also well known, as is shown in the images on the far left of a denuded hillside. The challenge that united idealists and materialists was how to protect the world both from and for its dominant species.

The world's first national park was established in California in 1864; by the 1920s such parks had been opened on every continent. The preservation of the forests and grassland within these regions maintains soil productivity and the supply of water for urban reservoirs. Parks also represent a considerable recreational and aesthetic value, and often yield sizable revenues from tourism. But probably their most important benefit is the protection that these sanctuaries afford to a still unknown number of plants and animals whose life cycles and food chains are essential to the continued existence of all forms of life, including humans.

Elephants range across the grassland of Kenya's Tsavo National Park, prey for tourists' cameras rather than poachers' guns.

were shipped east to be ground up for fertilizer. By the end of the century there were perhaps 500 surviving plains buffaloes in the United States.

Against destruction on such a massive scale, Thoreau's angry cry of protest could by itself achieve little. But he was not alone. In Europe at the beginning of the century a number of artists and writers, repelled by the general squalor and filth of the cities, had begun to look to the natural world not just as a material resource but as the font of spiritual wisdom. They became known as the Romantics, members of a cultural movement that caused a significant change in attitudes towards untamed nature. The English poet William Wordsworth, who was born in the rugged county of Cumberland, derived "sermons in stone" from his native Lakeland mountains. His French contemporary François de Chateaubriand, a self-styled "solitary wanderer", noted that "it is in the prospect of the sublime scenes of nature that the unknown being manifests himself to the human heart."

This Romantic love of nature soon captured the imagination of Europe's wealthy citizens. Trips to the Alps or to the Highlands of Scotland became a popular tonic for those who were distressed by the brutish scenes of industry all around them. More practically, steps were gradually taken to curb some of society's most obvious atrocities. The English philosopher Jeremy Bentham gave voice to a general new sensitivity towards animals: "The question is not can they reason, nor can they talk, but can they suffer?" The official response was that at least some of them could. In 1822 the British parliament passed an act which forbade cruelty to the largest domestic animals (dogs, cats and birds were not included, however). Two years later the Society for the Prevention of Cruelty to Animals became the first such national organization in the world. France established a similar society in 1845, followed shortly by most other European countries.

Meanwhile, observant members of the European colonial administrations out in the tropics and elsewhere had begun to voice severe warnings about the effects of rampant imperialism on their indigenous flora and fauna. These effects were most apparent in isolated oceanic islands such as Mauritius and St. Helena, which provided what amounted to laboratory conditions in which to note the depletion of native plants and animals that was caused by deforestation and overgrazing. As a result, practical measures to combat soil erosion and other forms of despoliation were taken in the Canary Islands, the West Indies, Mauritius and other islands.

In India the medical service of the British East India Company compiled a report in 1852 on the subject of "the probable effects in an economic and physical point of view of the destruction of tropical forests". The authors pointed to extensive soil erosion and the silting up of harbours that had occured along the Malabar Coast, and argued that large-scale deforestation could bring about a decline in rainfall and, eventually, famine. Serious droughts in the 1830s and later in the 1860s and 1870s appeared to bear these suppositions out, and in their wake programmes designed to protect the forests were established. Clearly, where it could be proved to the authorities that environmental degradation was not in their economic interests, and where the spectres of agricultural failure and social unrest were placed before their eyes, they were prepared to listen.

In America, too, when the voice of cool scientific reason was added to Thoreau's urgent, passionate pleas, both the government and the general public sat up and took notice. In 1864 George Perkins Marsh, America's first ambassador to the newly unified kingdom of Italy, published a book entitled *Man and Nature* in which he argued that the preservation of the wilderness served both "economical" and "poetical" ends. Forests were not useless, he pointed out. They preserved precious water resources and prevented erosion of the soil. Above all he cautioned against ignorant and unthinking assaults on the environment, declaring that "We can never know how wide a circle of disturbance we produce in the harmonies of nature when we throw the smallest pebble into the ocean of organic life."

The theories of experts in other fields also provided ammunition for the preservationists. Most humbling of all were those developed by Charles Darwin in 1859 in his book *On the Origin of Species*, which put an end to the notion that men and women occupied a unique and specially privileged position on earth. Darwin persuasively argued that the human race, for all its superior powers of invention, was not too distant a relative of the great apes.

The German biologist Ernst Haeckel developed Darwin's ideas a stage further. In 1869 he coined the world "oecology" (the "o" was dropped in the 20th century) to describe the intimate relationship between all living things. Haeckel was proposing that human beings were not merely cousins of the apes, they were working partners of the flea, the algae and the rat besides. In a way this concept was a modern restatement of the medieval idea of a great chain of being, which also posited links between humans and all other species inhabiting the earth.

Another tenet of traditional knowledge received scientific confirmation at about the same time. "For soil to remain fertile," wrote the German chemist Justus von Liebig in the 1840s, "that which has been taken from it has to be replaced completely." This theory had been blindingly obvious to countless generations of farmers in previous societies, but Liebig now proved it scientifically by analysing the chemical composition of plants and discovering which substances were removed from the soil when a crop was harvested. Phosphorus, potassium, calcium, magnesium and nitrogen, he discovered, were the elements most quickly lost when land was cleared without any return. Liebig advocated the use of human manure as a means of revitalizing the soil (and incidentally solving the industrial city's waste disposal problem), a practice that had been followed for centuries in China and certain other regions of Asia. When this suggestion failed to take hold he developed mineral nutrients that would allow the same patch of ground to produce crops year after year. The chemical fertilizer industry, which Liebig and others helped inspire, was to become the mainstay of 20th-century agriculture.

Even those who prized the land only for what they were able to get out of it now began to understand that they had to give as well as take, and that it was in their own interests to help protect this precious resource. In 1864 the United States government granted the spectacular Yosemite Valley to the state of California as a public park. Eight years later, in a dramatic concession to the nation's general change in mood, more than 800,000 hectares in the northwest of Wyoming became the Yellowstone National Park, the first acknowledgement by any government of the necessity of the large-scale protection of man's natural environment. It was a victory for the preservationists, even if only a qualified one. It was not for another 22 years, in 1894, that hunting was finally prohibited in the park, by which time only 200 of Yellowstone's once vast buffalo population remained alive.

The latter provision resulted from the energetic support that was given to the preservationists by America's sportsmen and big-game hunters. This was not, however, such an unlikely alliance as it might at first seem. Many hunters were often keen and

The rusting hulks of fishing boats lie beached on salt-encrusted wasteland once covered by the Aral Sea in the central Asian region of the U.S.S.R. The destruction of the local fishing industry is just one result of a failure to assess the environmental impact of drawing off the two major rivers that feed into the sea, the Amu Darya and the Syr Darya, to irrigate land for growing cotton. Between 1960 and 1989 the surface area of the sea has shrunk by more than 40 per cent *(inset map)*. Summer temperatures soar to 46 degrees Celsius, dust storms haze the sky, and pesticides and defoliants from the cotton fields contaminate drinking water. The local incidence of throat cancer and respiratory and eye diseases has increased, and the infant mortality rate is four times the national average.

knowledgeable naturalists, and vice versa. During the 1870s amateur hunters grew increasingly disturbed by the activities of the professionals, whose interests were solely commercial. They drove deer into the water with dogs, then clubbed them to death from boats; they baited and trapped wild turkeys, and slaughtered birds by the million for their plumage. And now the plains buffalo, a potent symbol of the frontier, was on the verge of extinction. It was all too much for the sportsmen, whose most outspoken champion was the future U.S. president Theodore Roosevelt. The Boone and Crockett Club, which Roosevelt helped found in 1888, pledged that it would "work for the preservation of the large game of this country", as well as "promote

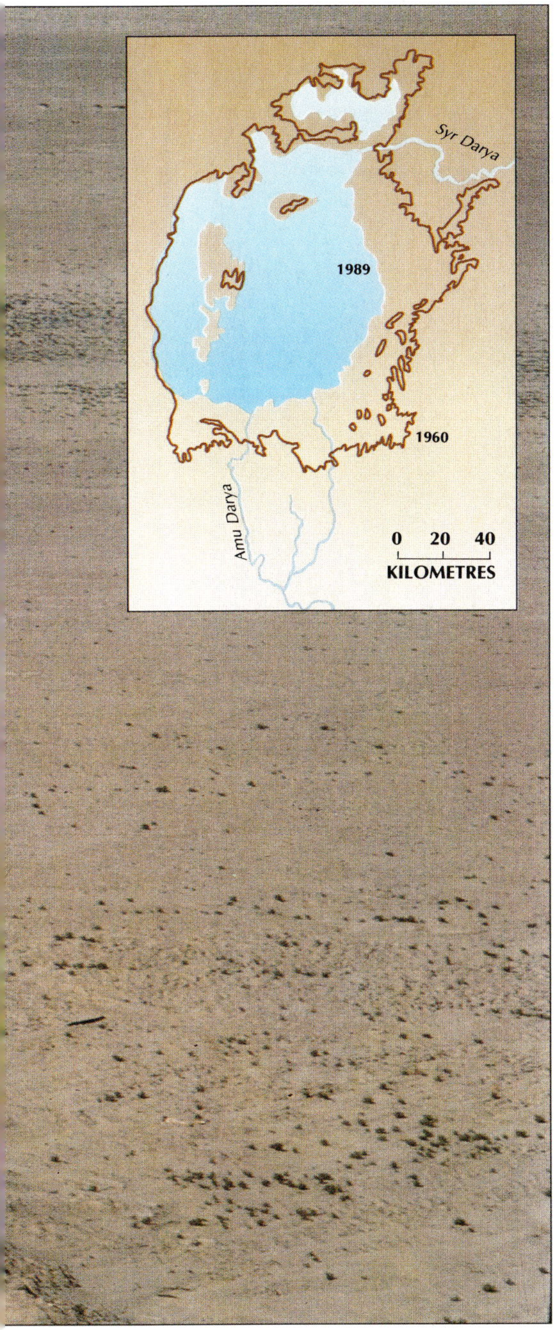


manly sport with the rifle". During his two terms in office from 1901 to 1909, President Roosevelt oversaw the establishment of five national parks, a professional forest service and 40 million hectares of forest reserve.

Not surprisingly, the material value of this land was the chief argument for its protection. Most conservationists had very little sympathy with the spiritual concerns of Thoreau or the Romantic poets. A forest should be preserved not because it was beautiful, but because it helped the soil retain water and, if properly managed, would serve as a steady source of timber. The attitude of President Roosevelt's chief forester Gifford Pinchot was strictly businesslike. "The fundamental principle of the whole conservation policy is that of use," he argued, "to take every part of the land and its resources and put it to use that will serve the most people." And it was this philosophy, rather than that of the idealists, that appealed to the policy makers of ensuing decades. Pinchot's words were to be precisely echoed in the terms used by the United Nations in 1969 to define conservation: "the rational use of the environment to achieve the highest quality of living for mankind".

For at least the first half of the 20th century the leaders of the world's developed countries were chiefly concerned with either preparing for war or recovering from its effects, and issues that related to the environment ranked relatively low on their scale of priorities. A number of isolated gestures were made: Belgium's king Albert established Albert National Park in the Congo in 1925, giving the gorilla much-needed protection. In the following year the government of South Africa, with an eye to the potential profits that could be made from tourism, designated 20,000 square kilometres as the Kruger National Park. But a promising international Conference for the Protection of Nature which assembled in Switzerland in 1913 fell into disarray with the advent of World War I. A similar fate attended any attempts made at international cooperation during the 1930s – all smashed into oblivion by the most destructive war that the world had ever witnessed.

Meanwhile, the earth's population continued to grow and the areas of wilderness to diminish. Improvements in sanitation and medicine in the developed world resulted in a significant decline in deaths from disease, particularly among children, and between 1850 and 1930 the number of people on the planet doubled from one to two billion. By 1976, when scientific advances extended to the countries of Asia and Africa, the population had doubled again. To provide these extra mouths with food, an estimated 432 million hectares were converted to arable land between 1860 and 1920; a further 419 million hectares were put to the plough during the succeeding 60 years. The inevitable overall result of this rapid, unchecked expansion was the accelerating depletion of the natural world's living resources.

The application of industrial technology to agriculture freed many farmers from backbreaking manual labour, and improved production. Following the introduction of the small petrol-driven tractor in the 1920s, for example, the number of hours worked by a modern farmer on a given area of wheat was reduced to one sixth of the time that his great-grandfather had needed. But the increasing intensity with which land was farmed could have unlooked-for results, especially when the soil was used for the continuous growing of one type of crop. Golden fields of ripe wheat may give the impression of abundance to hungry human beings, but in reality they are an ecological desert: very few other species can live in an environment devoted exclusively to the production of a single crop. Even irrigation, intended to bring life into

desert places, often has the opposite effect. Where drainage is inadequate, the land can quickly become waterlogged and useless. Salt accumulates rapidly in irrigated regions, either from evaporation or the dissolution of solid salts present in the earth. Because of waterlogging or salinity, irrigated land is in fact abandoned at about the same rate as it is brought into service.

The environment's response to overexploitation was nowhere more dramatic than in the southern regions of America's Great Plains. In their virgin state the dry plains of western Kansas, Oklahoma and northern Texas were covered with a tough, complex carpet of plant life. Up to 75 species of short grasses alone might occupy any one area. Even overgrazing by cattle in the late 19th century had not destroyed this well-adapted community of vegetation. But the self-styled "sod busters" of the early 20th century were a different matter. There were good profits to be made from the nation's demand for wheat and, so long as prices remained high, the plains farmers went on busting the sod. By 1925, in southwestern Kansas, there were some 800,000 hectares under cultivation; half as much again was torn up over the next five years.

Then disaster struck. From 1931 to 1936 yearly precipitation fell from 48 centimetres to just under 30. Crops failed over the five worst-affected states. In 1937, 100,000 hectares were planted with wheat in Cimarron County, Oklahoma; none were harvested. Lashed by spring winds, the bare, parched soil flew up into the sky. Brown snow fell in New Hampshire; dust settled on ships at sea.

The government responded with the best means it had. Soil conservation experts descended upon the Dust Bowl region to help those farmers who had not already fled west. Erosion could be discouraged by the terracing of slopes, the experts advised, or by ploughing along the contours of the land rather than in rigid straight lines. Trees should be planted as windbreaks, and some land should be kept grassed over. For a while, at least, the chastened farmers paid attention to what the experts had to say, but when the rains resumed and another war broke out in the 1940s, the urge to profit from the land at any cost overwhelmed their better judgment. "What we are doing in the Great Plains today is nothing short of soil murder and financial suicide," admitted the United States secretary of agriculture in 1947, but the murder continued unabated. And when the rains failed, dust-bowl conditions returned to the region in the mid-1950s, and again in the mid-1970s.

**During April 1985, refugees from the famine-stricken Ethiopian province of Tigré – where three rainless years have left the earth parched and cracked – trek towards hoped-for food and medical relief in the Sudan. In Ethiopia, as elsewhere, famine is a crisis not so much of food shortage and starvation as of political mismanagement. While the governments of some of the world's poorest nations, such as Botswana and Costa Rica, have managed to provide their citizens with sufficient food, others exploit inequitable systems of distribution: the Ethiopian regime in the late 1980s bombed 25,000 tonnes of food aid to prevent it being eaten by civilians not under government control. Most famine deaths occur from the spread of disease that follows social breakdown and mass migration, which in Ethiopia has included the forced resettlement of people from drought-stricken areas.**

Increasingly, both farmers and conservationists looked to science to alleviate ecological damage while at the same time augmenting food production. At first it seemed as if these two apparently irreconcilable demands could be met. New strains of high-yield food crops, fed with chemical fertilizers and protected from predators and weeds by a fearsome battery of recently developed pesticides, dramatically increased the crop yields in Western countries and in the poorer nations of the Third World. Nuclear power, which had been adapted for commercial purposes after its development for military use in the Second World War, promised a pollution-free alternative to fossil fuels. But just as satisfied customers across the world were beginning to take the fruits of these developments for granted, the disastrous scale of their destructive side effects became apparent.

Pesticides gave cause for the first cry of alarm. "As crude a weapon as the cave man's club, the chemical barrage has been hurled against the fabric of life," wrote the American biologist Rachel Carson in *Silent Spring*, a passionate attack on the pesticide industry. Writing in 1962, she conjured up the image of a world without songbirds, and indeed birds were falling victim to chlorinated hydrocarbons such as

# THE HARVEST OF TECHNOLOGY

Scientists examine hybrid strains of rice at the International Rice Research Institute in the Philippines, established with grants from U.S. foundations in 1962. The institute's first major success was a crossbred variety of rice with a yield three times that of traditional plants, which by 1968 was grown in 80 countries.

Seedlings developed by tissue culture – which involves the rearing of cells isolated from living plants – are suspended in test tubes in a sterile growing medium which contains hormones for their regeneration. By using the parts of a plant uncontaminated by disease, tissue culture can produce virus-free crops. It is also a preparatory stage in plant genetic engineering, in which genes are inserted into single cells.

Twentieth-century biochemistry and genetics have enabled the development of fast-maturing plants that allow more crops to be grown each year, and the breeding in of genetic resistance to pests and diseases and to harsh conditions such as drought and salinity. In the 1960s, when the effects of this application of science to crop growing became known as "the green revolution", many of the new plants required good soil and plentiful amounts of water and fertilizer to produce their increased yields; since then, further advances in biotechnology have increased the world's food production by 50 billion tonnes per year. The benefits have been most apparent in parts of Asia and South America, where food supplies have more than kept pace with the population increase; in Africa, progress has so far been slowed by various political obstacles and local land tenure systems. But the uses of biotechnology are not always beneficial to Third World countries: the laboratory production of substances such as cocoa butter in countries that previously had to import them has the potential to transform world markets – to the probable advantage of the richer nations.

Lettuces flourish in a commercial greenhouse without soil, grown by a technique known as hydroponics. The plant roots are immersed in a water solution containing the minerals necessary for growth, or in gravel or sand to which a nutrient solution is added. This method of cultivation is appropriate for regions that have a favourable climate but are lacking in fertile soil, but installation costs are high.

DDT, which accumulate in the bodies of any organism that consumes them. Birds high up in the food chain contained the highest concentration of pesticides: their diet of smaller birds, mammals or fish was already loaded with a deadly chemical cocktail. Fish in Lake Michigan were found to contain 3 to 8 parts per million DDT in the 1960s; in herring gulls, which ate the fish, this figure rose to 3,177. Birds of prey laid eggs with shells so thin they broke in the nest.

While their natural predators died, many species of agricultural pest developed a resistance to the poisons directed against them, forcing scientists to develop yet another chemical weapon, a process that became known as the "pesticide treadmill". One variety of malarial mosquito that appeared in the late 1980s was immune to four of science's most deadly pesticides: DDT, aldrin, dieldrin and Malathion.

Many countries proceeded to ban all these poisons, but other forms of environmental pollution proved rather more difficult to legislate against. One such problem was the runoff from chemical fertilizers. Adding water-soluble nutrients to the soil increases crop yields, but unfortunately the crop does not get the full benefit of these chemicals. Excess nitrates and phosphates are washed away into streams, lakes or ground water. Water enriched by nutrients such as those that come from fertilizers becomes extremely attractive to plant life, particularly to blue-green algae, which has a tendency to multiply rapidly at the expense of other aquatic life. A dense growth of algae uses up oxygen at such a rate that fish and other plants are unable to compete. This and other forms of pollution have caused a steep decline in commercial fishing on Europe's inland waterways.

On land, chemical fertilizers can have an even more serious long-term effect. Lakes are surprisingly resilient to chemical pollution, and often in time recover their natural balance. The soil may not. A complex mixture of minerals, microorganisms and vegetable matter and soil is anchored by the roots of plants, which both feed it and feed from it. Dressing a field with manure or vegetable compost fertilizes the soil and also helps to bind together its mineral components. Chemical fertilizers ensure that crops will grow, but provide none of the sustaining body. Deprived of its plant cover and its natural humus, the crumbly topsoil begins to blow or wash away. In 1977 it was reported that American farmers were losing 2.5 centimetres of topsoil every 16 years – an amount that might be renewed naturally only every 300 to 1,000 years. As a result, one third of American agricultural land was judged to be losing productivity.

The dilemmas posed by the use of science and technology to increase food production were not only economic – in terms of balancing short-term gains against long-term losses – but also moral. The latter were especially evident in the case of animals, whose treatment increasingly bore out the law of progress observed by the Irish poet Oliver Goldsmith in the 18th century: "In all countries, as man is civilized and improved, the lower ranks of animals are oppressed and degraded."

For centuries those who believed that meat was a natural part of the human diet had scarcely been troubled by ethical concerns. Advocates of vegetarianism had argued that, in the words of one 17th-century English writer, it was "unlawful to kill any creature that had life, because it came from God" – but they had attracted few followers. With the introduction of intensive farming methods during the late 20th century, however, the degradation endured by animals reached such a scale that it made the moral issues hard to avoid.

Factory farming in Western countries turned poultry, pigs and, to a lesser extent, cattle into utilitarian objects, raw material for the production of meat. On most intensive farms, piglets are removed from their mothers within hours of birth. Without anaesthesia, their incisors are clipped and their tails docked, and the males are castrated. Housed in cages with metal-slatted or cement floors, they are kept in darkness and at a constant humid temperature – to induce lethargy and so that food is not used up in keeping them warm. They are injected with a range of antibiotics and have growth-promoting agents added to their feed. They are slaughtered at six to eight months of age. Such treatment may be rationally justified on economic grounds: it yields the maximum amount of meat at the minimum cost. But neither reason nor economics can explain why certain other animal species are treated in an entirely different manner. In the countries of the European Economic Community in the early 1980s, affection was lavished on an estimated 91 million dogs, cats and other pets. In the United States, families spent an astonishing $7.5 billion per year on food and veterinary care for an estimated 475 million household pets.

At a purely personal level it may be quite possible to resolve such inconsistencies in differing cultural attitudes towards nature – by becoming vegetarian, perhaps, or by buying produce exclusively from the small number of farmers who have reverted to traditional methods in growing their crops or raising their animals. At a global level the choices are not that easy. The natural world is in fact eroded more by human need than it is by greed, and in the final quarter of the 20th century the scale of that need has continued to accelerate.

In the 14 years from 1976 to 1990 the number of people on earth grew by about one billion, bringing the total world population to slightly over five billion. Ninety per cent of this growth occurred among the developing countries, which had begun to reap the benefits of modern medicine and the fast returns of chemical agriculture, while still lacking the capacity, and often even the desire, to adopt an efficient programme of population control. The fastest growing region of all was Africa: between 1960 and 1988 the population of that continent more than doubled, rising from 280 million to over 620 million. During the same period, Asia's population increased from nearly 1.7 billion to a staggering three billion. Although different forecasts vary, it is predicted by many that the earth's total population will double again to around 10 billion in only another 40 years.

In China and in certain other parts of Asia where the rate of population growth actually began to slow, food output more than kept pace with demand. In Africa, on the other hand, where agriculture suffered from a labour shortage because of a drift to work in the mines and cities, yields were low and variable. And in many countries of the Third World the pressure on land had tragic consequences. The droughts that struck sub-Saharan Africa during the mid-1980s, killing many hundreds of thousands of people, would have had far less impact had the land not already been overgrazed. In the impoverished and densely populated country of Bangladesh, families migrated out on to rich new lands that had been created by silt washed down from the hills. Nothing there protected them against floods from the mountains or murderous inundations from the ocean. In 1970 a tidal wave drowned an estimated 300,000 Bangladeshis on the Brahmaputra-Ganges delta. Scarcely had the sea receded than thousands more of their countrymen descended to take their place, risking the inevitable onset of future natural disasters.

With pressure such as this, the preservation of the wilderness was not a luxury many countries could seriously contemplate. Towards the end of 1970s the Brazilian

**In the California desert, 1,818 mirrors seven metres square reflect sunlight on to a central receiver; steam from water pumped through the receiver drove a turbine producing 10 megawatts of power, eight hours a day. This experimental field was shut down in the late 1980s, but similar projects continued in Spain, Israel and New Mexico. Apart from tidal power and geothermal heat from the earth's core, all forms of renewable energy derive ultimately from the sun. The form in which this energy is currently most widely exploited is hydropower: sunlight evaporates water, which later falls as rain and flows into rivers which can be used to drive electricity turbines. In 1987, 17 per cent of the electricity used in industrial countries and 31 per cent in developing nations was produced by hydropower.**

government launched a determined campaign to use its "waste" land for the needs of its expanding population. Rainforests are in fact the world's richest repositories of life, believed to contain between 50 and 80 per cent of all plants and animals, and the rainforests of the Amazon Basin comprise the largest remaining system, covering an area equivalent to almost 90 per cent of the United States. For the government of Brazil, however, the country's need for agricultural land, and for the hard currency to be derived from exports of beef, overrode international outrage. Cattle ranchers and subsistence farmers advanced along new highways, clearing the land by fire and decimating the native population of Amerindians. In 1990 United Nations experts

determined from looking at satellite photographs that as much as 20 million hectares of precious rainforest were being destroyed by fire annually.

Other wilderness areas in other parts of the world had their own attendant problems. The East African nation of Kenya has an annual population growth rate of 4 per cent, the world's highest, yet here, by a twist of irony, economic arguments actually favoured the preservation of protected game reserves rather than their conversion to agricultural use. In 1977 an adult male lion in Kenya's Amboseli National Park was worth an estimated $1,150 to a poacher; to the government its value in tourist revenue was around $515,000. As a result of this persuasive lesson in economics Kenya banned all sports hunting and attempted to save its endangered elephants by waging a shoot-to-kill war against ivory poachers. But hungry people in need of land for crops may yet do more damage to Africa's wildlife than guns ever have.

Elsewhere even those wilderness areas that are too inhospitable for human habitation came under threat. As the world petroleum and uranium reserves grew scarcer, many believed that "limited" exploration for natural resources should be permitted in the great virgin landmass of Antarctica. Others argued that humanity is no great respecter of limits and opposed any development there whatsoever. They pointed to the case of Alaska, often described as "America's last frontier". When huge reserves of oil were discovered at Prudhoe Bay in 1968, oil companies built a 1,300-kilometre overland pipeline snaking across the state to the town of Valdez lying on Alaska's south coast. In 1989 the tanker *Exxon Valdez* ran aground on rocks in Prince William Sound, blackening 2,000 kilometres of Alaska's coastline and destroying the local fishing industry together with an estimated 400,000 sea birds. Six months after the spill and a cleanup operation involving 11,000 workers, around 60 per cent of the oil still remained on the once pristine shore.

In the 1980s it became apparent that not even Europe was immune: the woods and lakes that comprise that continent's own dwindling wilderness were dying from an industrial phenomenon known as acid rain. Sulphur and nitrogen oxides, released from car exhausts or erupting from factory chimneys, dissolve in water to form dilute solutions of sulphuric and nitric acid. This descends as rain or snow, often hundreds of kilometres from the source of emission. In 1982 it was thought that only 8 per cent of German forests were affected; two years later the figure had risen as high as 50 per cent and the problem had been discovered to exist throughout Europe.

Acid rain brought home to many Europeans the fact that large-scale changes wrought on the environment by technology and the economic policies of the world's richest nations took effect not only in exotic faraway places, but in their own back yard. For Americans and for Eastern as well as Western Europeans, this message was reinforced by leaks of radioactivity from nuclear power stations at Three Mile Island in Pennsylvania in 1979 and at Chernobyl in the Soviet Union in 1986. And the rider to these accidents was that even a change to an alternative form of energy from those on which Western prosperity had been founded was no guarantee of security.

By the final decade of the century environmental issues had advanced to the top of the political agenda. The dire predictions of experts gathered at international conferences were increasingly brought to public attention, and governments in the Western world – which were elected by majority vote – were forced to sit up and take notice. The word and the colour "green" became a media symbol, loaded with political significance. Manufacturers and advertisers, responding to new regulations and with an eye to future market trends, created a fashion for products that were claimed to be environmentally "safe". When motorists stopped at a garage they were offered unleaded petrol, which most understood to have a less polluting effect than the petrol they had previously bought. In many neighbourhoods bins were provided in which glass bottles and aluminium cans could be deposited for the recycling of these materials. In supermarkets, certain aerosol spray cans were labelled as "ozone friendly". This simplified message, combining science and sentiment, in fact meant that they contained no CFCs, or chlorofluorocarbons, which attack and destroy atmospheric ozone, the gas which prevents excessive amounts of harmful ultraviolet light from reaching the earth. In the circumstances, however, it was not necessary to understand scientific jargon to know that a crisis was at hand.

Nor did consumers have to study scholarly journals to know what this crisis was about. They could read in the newspapers of new diseases transmitted by the food people ate, food that had been eaten safely for centuries. The "greenhouse effect", they were told, which had to do with global warming from the burning of fossil fuels and the prevention of the sun's infrared heat from leaving the atmosphere, was increasing temperatures and would cause a rise in sea levels and the flooding of low-lying areas. Such things would occur, if not in their own lifetimes, then in those of their children or grandchildren. A growing number of people began to realize that what was at stake was not just the rainforests or the endangered species featured on television documentaries, but many of the things that gave their own lives meaning, and which they had so far taken for granted: clean water, fresh food, the parks and the open spaces where they took their families at weekends.

As in all times of crisis, it was often difficult to keep a clear sense of perspective. The grand sweep of history appeared to provide little comfort: the decline of Rome and many other empires was caused at least in part by mismanagement and over-exploitation of their natural resources. But another factor in their decline was often their succumbing to a mood of inevitable doom; and on closer inspection the narrative of human impact on the natural world offers at least as many positive as negative lessons. In the 19th century the North American bison was hunted to the point of extinction; on the other hand, timely human intervention in the form of national parks and conservation laws ensured its survival. In the present century the use of chemical pesticides caused immense harm to many birds and animals, but strict legislation in Western countries reduced levels of DDT in the environment to almost zero. In the late 1980s clearance of the Amazonian rainforests increased at an alarming rate – but the bulk of the lowland forests still remains intact, and humans now possess the knowledge and the means to conserve almost all species, as well as to inhibit land degradation and return rainfall to the atmosphere.

Undoubtedly the economic and political policies of the world's richest nations, while accruing enormous material benefits for their peoples, have caused damage to the natural world. By the 1990s, however, this fact was in the Western world more widely acknowledged than ever before. There was also an increased awareness that on a global scale the maintenance of economic and ecological health amounts to the same thing. The problem facing political leaders was how to translate this awareness into action. They could take heart from earlier generations of environmentalists, who reminded their audiences that humans are not passive victims of deterministic forces, but have the power to influence their own destiny. "It is never too late to give up our prejudices," advised Thoreau. "What old people say you cannot do, you try and find that you can. Old deeds for old people and new deeds for new."

**A colour-enhanced computer image of the southern hemisphere, generated from data collected by a U.S. satellite in October 1989, shows a black "hole" in the stratospheric layer of ozone that filters the sun's ultraviolet rays. This radiation can increase the incidence of skin cancers and immune deficiencies, and may damage crops and aquatic ecosystems. The inert gases that destroy ozone – chiefly CFCs, used in aerosols and refrigerators – were unilaterally banned by some Western nations in 1978, and timetables for their phasing out were agreed by other countries in the 1980s. These decisions were prompted by scientists' new discoveries about the interaction of humans and their planet, relayed by political pressure groups such as those whose insignias are shown above.**

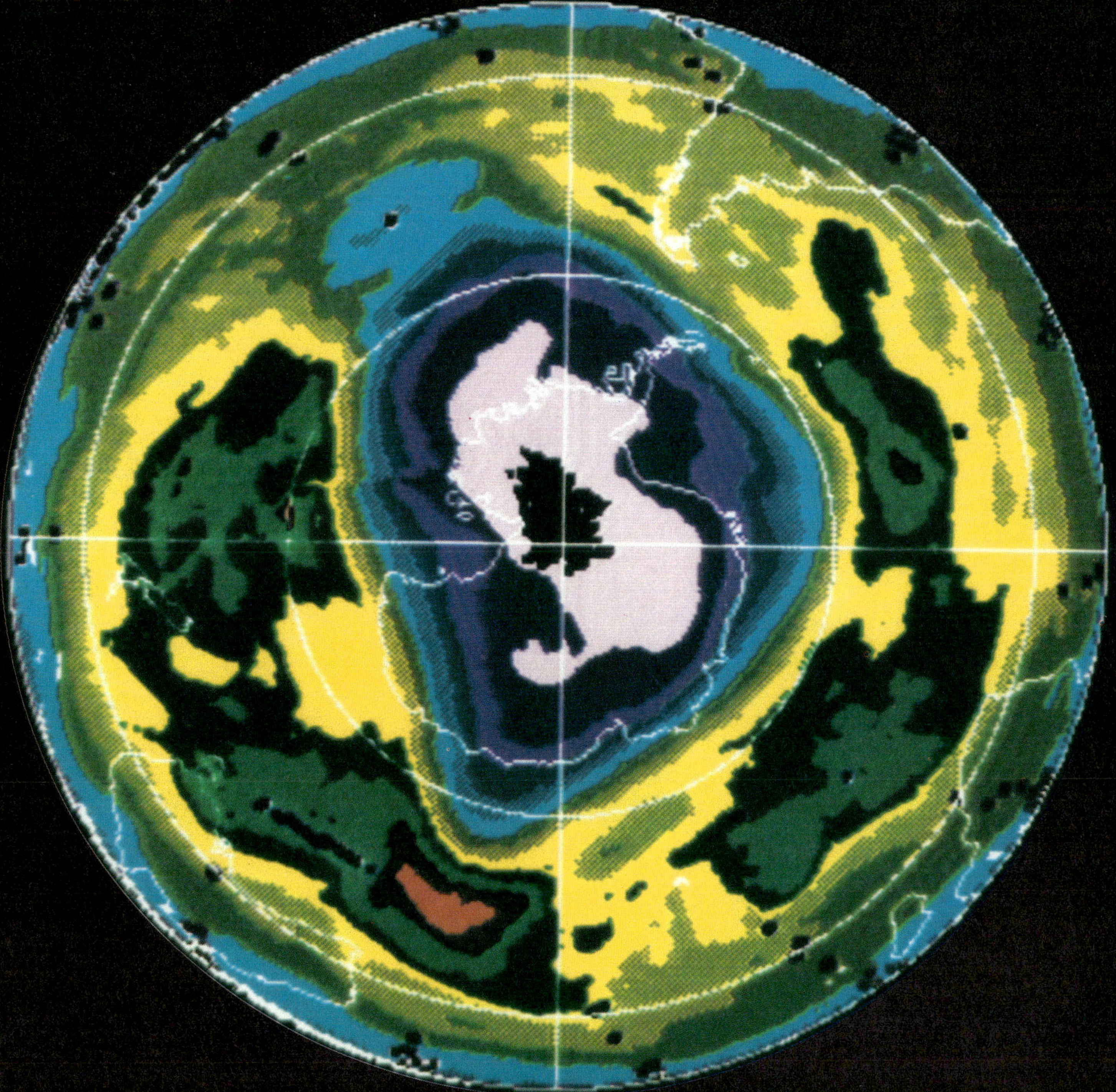

## 100,000-10,000 BC

Around 100,000 BC the first fully modern humans *(Homo sapiens sapiens)* emerge, probably in eastern and southern Africa, living by hunting animals and gathering plants.

Bands of human hunter-gatherers move into Eurasia from Africa (40,000 BC), replacing pre-existing Neanderthals, and cross the sea gap from Southeast Asia into New Guinea and Australia.

From around 20,000 BC the use of new weapons – bows and arrows, spear-throwers – and tactics such as ambushes and the use of fire to stampede herds increase the hunting efficiency of early humans.

Paintings of animals are executed in caves in southwestern France and northern Spain (15,000 BC).

Grindstones are used to make flour from wild grass seed in Upper Egypt (12,000 BC).

Hunter-gatherer bands reach the tip of South America (10,000 BC).

Climatic change and the scale of human hunting contribute to the extinction of species such as the mammoth, woolly rhino and the American horse.

**10 million**

## 10,000 BC-0

The cultivation of domesticated wheat, barley and pulses, and the herding of livestock (cattle, sheep, goats, pigs) spread outwards from the Fertile Crescent (8000 BC). Agriculture develops independently in Southeast Asia and the Americas.

Widespread sea and land trade in precious stones and shells develops (7000 BC).

North African fishing and cattle cultures flourish before the drying-out of the Sahara (4000 BC).

Cities, founded on agricultural prosperity, are built in Mesopotamia (3500 BC). Shortly afterwards, urban centres are founded in Egypt.

Trade and agriculture bring wealth to the palaces of Minoan Crete and Mycenaean Greece (2000 BC).

Horses are domesticated; pastoral nomads range across the Asian steppes with flocks of sheep and goats (1500 BC).

Cities supported by rice growing are established in India's Ganges Valley (800 BC).

Greek and Phoenician colonies are established around the Mediterranean (750 BC); Athens and other city-states flourish on the Greek mainland (fifth century BC).

The Roman republic is established (550 BC).

The Chinese empire is unified under the Qin dynasty (221 BC).

Rome becomes dependent on the import of grain from Sicily, Sardinia and, after the destruction of Carthage in 146 BC, North Africa.

**200 million**

## 0-AD 1000

The growing population of Rome, reaching one million in the first century AD, spurs further military conquests. By the second century the Roman empire extends from Spain and Britain in the west to Syria in the east.

Tropical forest civilizations develop in lowland Mexico and Southeast Asia.

The Bantu peoples spread across sub-Saharan Africa.

Paper is manufactured in China (second century).

Rome is sacked by the Visigoths (410).

Mesopotamian irrigation systems help support a growing population (500-800).

Arab armies spread the Islamic religion through the Middle East and North Africa (seventh century); through the medium of Arab scholars, the classical learning of Greece and Rome is later passed on to Europe.

In China the Grand Canal is constructed between the Yangtze and Yellow rivers to transport grain from the south to the north (seventh century).

Viking seafarers settle in Iceland (870) and Greenland.

The Polynesians, having colonized most oceanic islands in the Pacific, settle in New Zealand (ninth century).

**300 million**

## AD 1000-1500

In Europe, favourable climatic conditions from the 11th century promote advances in agriculture and an increase in population. Crop rotation and open fields divided into strips for efficient ploughing are introduced. Wine from the south is traded for grain from the north.

In China, the introduction in the 11th century of a fast-maturing rice from Vietnam allows more than one harvest per year. China also benefits from improved technology such as lock gates in canals, machines for textile production and the invention of movable-type printing.

Windmills are in use in Europe from 1180.

In the highlands of Central and South America, Aztec and Inca civilizations reach their apogee (1300-1500).

The Black Death, beginning in Asia, reaches Europe in 1347 and kills more than a third of the continent's population.

African sub-Saharan civilizations flourish in Zimbabwe (14th century) and West Africa (Ghana, Songhai).

Open fields in Europe are enclosed for sheep farming from the 15th century.

Johann Gutenberg produces the first printed books in Europe (1455).

The voyages of European navigators prepare the way for the global expansion of European power: Bartolomeu Dias sails round the Cape of Good Hope (1488); Christopher Columbus crosses the Atlantic and makes landfall in the Caribbean (1492); Vasco da Gama reaches India (1498).

**400 million**

## AD 1500-1990

The transportation of slaves from West Africa to the Americas begins (1505).

Spanish conquistadors establish a colonial empire in Central and South America (1519-1535); European diseases kill off 90 per cent of the indigenous populations.

The potato is imported to Europe from South America (1525). Other imports from the New World include maize, tobacco and the turkey. Coffee and tea are introduced to Europe from the Orient.

The first permanent English settlement in North America is established (1607), shortly followed by French and Dutch settlements.

The "Little Ice Age", beginning around 1590, causes failed harvests and the abandonment of uplands in Europe.

The introduction of seed drills and similar technology in the early 18th century signals the beginning of the mechanization of European agriculture.

Steam-engines are developed in Britain for use in coal mines (1765).

The establishment of a penal colony at Sydney Cove in 1788 inaugurates the British settlement of Australia.

Steam-powered trains are introduced in Britain (1825); extensive railway networks are built in Europe and America, and subsequently in Australia, Africa, India, Japan and Russia.

The Industrial Revolution in Europe and later in America promotes the growth of towns and accelerates the impact of human technology on the natural world. From the 1830s the mechanization of Western agriculture is speeded by the introduction of the mechanical reaper, the steam plough and the combine harvester.

The first oil well begins operation in Pennsylvania (1859).

Following advocacy for conservation, Yosemite Valley in California, U.S., becomes the first national park (1864).

Overcultivation of the American plains leads to drought and "Dust Bowl" conditions in the 1930s.

The use of chemical pesticides becomes widespread in Western agriculture in the 1940s and 1950s; strict legislation is enforced from the 1960s to limit their harmful side effects.

In China from 1949 the Communist government institutes radical programmes of collectivization and population control.

The first nuclear power station, offering an alternative energy source to fossil fuels, is opened in the U.K. (1956).

Genetic engineering is increasingly applied to crop growing and stockbreeding. The International Rice Research Institute is founded in the Philippines (1962) to develop new strains of high-yield rice.

The consequences of environmental deterioration in the 1980s – which include famine in Africa, global warming and the depletion of the earth's ozone layer – increase awareness in the West of the need for long-term management of natural resources.

**5,000 million +**

The chronology on the left lists some of the main events in the narrative of man's impact upon the natural world. The figures at the foot of each column represent the approximate human population of the world at the end of the given time span.

Population growth has always impressed itself on the landscape – by prompting migrations to previously uninhabited regions, by causing unused land to be put to the plough, or by spurring the development of new technologies. In the last two centuries, however, the curve has risen ever more steeply: the population increase in the past 40 years alone has equalled the total increase from the emergence of the human species up to 1950. Although the rate of increase is now gradually slowing, the absolute population figure continues to rise rapidly. The present population level is at once a measure of human success in deriving a living from the planet, and a challenge to the flexibility of political and economic structures.

# BIBLIOGRAPHY


**Anderson,** Perry, *Passages from Antiquity to Feudalism.* London: New Left Books, 1974.
**Anglicus,** Bartholomew, *Medieval Lore.* London: Elliot Stock, 1893.
**Arnold,** T.W., *The Legacy of Islam.* Oxford: Oxford University Press, 1961.
**Attenborough,** David, *The First Eden: The Mediterranean World and Man.* London: Collins/ BBC Books, 1987.
**Austin,** M.M., and P. Vidal-Naquet, *Economic and Social History of Ancient Greece: An Introduction.* London: B.T. Batsford, 1969.
**Aykroyd,** W.R., *Sweet Malefactor: Sugar, Slavery and Human Society.* London: Heinemann, 1967.
**Barker,** Graeme, *Prehistoric Farming in Europe.* Cambridge: Cambridge University Press, 1985.
**Bazarov,** Konstantin, *Landscape Painting.* London: Octopus Books, 1981.
**Berrall,** Julia S., *The Garden: An Illustrated History.* London: Penguin Books, 1978.
**Bloch,** Marc, *Feudal Society.* London: Routledge & Kegan Paul, 1989.
**Blum,** Jerome, ed., *Our Forgotten Past: Seven Centuries of Life on the Land.* London: Thames and Hudson, 1982.
**Blunt,** Wilfrid, *The Compleat Naturalist: A Life of Linnaeus.* London: Collins, 1984.
**Boardman,** John, Jasper Griffin and Oswyn Murray, eds.:
*Greece and the Hellenistic World.* Oxford: Oxford University Press, 1988.
*The Roman World.* Oxford: Oxford University Press, 1988.
**Braudel,** Fernand, *Capitalism and Material Life.* London: Weidenfeld & Nicolson, 1973.
**Brothwell,** D., and P. Brothwell, *Food in Antiquity.* London: Thames and Hudson, 1969.
**Browning,** Robert, ed., *The Greek World.* London: Thames and Hudson, 1985.
**Buchanan,** Keith, *The Transformation of the Chinese Earth.* London: G. Bell & Sons, 1970.
**Carson,** Rachel, *Silent Spring.* London: Hamish Hamilton, 1964.
**Coates,** Austin, *The Commerce in Rubber: The First 250 Years.* Singapore: Oxford University Press, 1987.
**Cressey,** George B., *Land of the 500 Million: A Geography of China.* New York: McGraw-Hill, 1955.
**Crosby,** Alfred W., *Ecological Imperialism: The Biological Expansion of Europe 900-1900.* Cambridge: Cambridge University Press, 1986.
**Cummins,** John, *The Hound and the Hawk: The Art of Medieval Hunting.* London: Weidenfeld & Nicolson, 1988.
**Cunliffe,** Barry, *Rome and Her Empire.* London: Bodley Head, 1978.
**Delano-Smith,** Catherine, *Western Mediterranean Europe.* London: Academic Press, 1979.
**Dilke,** O.A.W., *Greek and Roman Maps.* London: Thames and Hudson, 1985.
**Douglas,** Mary, *Purity and Danger: An Analysis of the Concepts of Pollution and Taboo.* London and New York: Ark Paperbacks, 1984.
**Duby,** Georges, *Rural Economy and Country Life in the Medieval West.* London: Edward Arnold, 1968.
**Edgerton,** Samuel, Jr., *The Renaissance Discovery of Linear Perspective.* New York: Harper & Row, 1978.
**Edlin,** H.L., *Man and Plants.* London: Aldus Books, 1967.
**Ehrlich,** Anne H., and Paul R. Ehrlich: *Earth.* London: Thames Methuen, 1987.
*Extinction.* New York: Random House, 1981.
**Elvin,** Mark, *The Pattern of the Chinese Past.* London: Methuen, 1973.
*The Epic of Gilgamesh.* Transl. by N.K. Sandars. London: Penguin Books, 1972.
**Fagan,** Brian M., *People of the Earth: An Introduction to World Prehistory.* Glenview, Ill.: Scott, Foresman, 1989.
**Farb,** Peter, *Humankind.* London: Triad Paladin, 1978.
**Galloway,** J.H., *The Sugar Cane Industry: An Historical Geography from its Origins to 1914.* Cambridge: Cambridge University Press, 1989.
**Gamble,** Clive, *The Palaeolithic Settlement of Europe.* Cambridge: Cambridge University Press, 1986.
**Garnsey,** Peter, *Famine and Food Supply in the Graeco-Roman World.* Cambridge: Cambridge University Press, 1988.
**Gimpel,** Jean, *The Medieval Machine: The Industrial Revolution of the Middle Ages.* London: Victor Gollancz, 1977.
**Glaber,** Radulfus, *The Five Books of Histories.* Ed. and transl. by John France. Oxford: Clarendon Press, 1989.
**Glacken,** Clarence, *Traces on the Rhodian Shore: Environmental Ideas from the Classical Period to the End of the Eighteenth Century.* Berkeley, Calif.: University of California Press, 1967.
**Goudie,** Andrew, *The Human Impact.* Oxford: Basil Blackwell, 1989.
**Gowlett,** John, *Ascent to Civilization.* London: Collins, 1984.
**Harris,** Marvin, *Cows, Pigs, War and Witches: The Riddles of Culture.* London: Hutchinson, 1975.
**Hartley,** Dorothy, *Lost Country Life.* New York: Pantheon, 1979.
**Hawkes,** David (transl.), *The Songs of the South: An Anthology of Ancient Chinese Poems.* London: Penguin Books, 1985.
**Healy,** John F., *Mining and Metallurgy in the Greek and Roman World.* London: Thames and Hudson, 1978.
**Hesiod,** *Theogony* and *Works and Days.* Transl. by Dorothea Wender. London: Penguin Books, 1973.
**Honour,** Hugh, *The New Golden Land: European Images of America.* London: Allen Lane, 1976.
**Hook,** Brian, ed., *The Cambridge Encyclopedia of China.* Cambridge: Cambridge University Press, 1982.
**Hunter,** J.M., *Land into Landscape.* London: George Godwin, 1985.
**Jenkins,** Ian, *Greek and Roman Life.* London: British Museum Publications, 1986.
**Jones,** E.L., *The European Miracle: Environments, Economies and Geo-politics in the History of Europe and Asia.* Cambridge: Cambridge University Press, 1987.
**Khazanov,** A.M., *Nomads and the Outside World.* Transl. by Julia Crookenden. Cambridge: Cambridge University Press, 1984.
**Klingender,** Francis, *Animals in Art and Thought to the End of the Middle Ages.* London: Routledge & Kegan Paul, 1971.
**Komarov,** Boris, *The Destruction of Nature in the Soviet Union.* London: Pluto Press, 1980.
**Ladurie,** Emanuel Le Roy, *Montaillou: Cathars and Catholics in a French*

*Village 1294-1324*. Transl. by Barbara Bray. London: Penguin Books, 1980.
**Lawson-Peebles,** Robert, *Landscape and Written Expression in Revolutionary America*. Cambridge: Cambridge University Press, 1988.
**Lee,** R.B., and I. DeVore, eds., *Kalahari Hunter-Gatherers: Studies of the !Kung San and their Neighbours*. Cambridge, Mass.: Harvard University Press, 1976.
**Lehner,** Ernst, and Johanna Lehner, *Folklore & Odysseys of Food & Medicinal Plants*. London: Harrap, 1973.
**Leroi-Gourhan,** A., *The Art of Prehistoric Man in Western Europe*. London: Thames and Hudson, 1968.
**Lewington,** Anna, *Plants for People*. London: Natural History Museum Publications, 1990.
**Loewe,** Michael, *The Pride that was China*. London: Sidgwick & Jackson, 1990.
**Longone,** Janice B., *Mother Maize and King Corn: The Persistence of Corn in the American Ethos*. Ann Arbor, Mich.: The William L. Clements Library, University of Michigan, 1986.
**Mackay,** Angus, *Spain in the Middle Ages*. London: Macmillan, 1977.
**McKeown,** Thomas, *The Modern Rise of Population*. London: Edward Arnold, 1976.
**McLuhan,** T.C., ed., *Touch the Earth: A Self-Portrait of Indian Existence*. New York: Simon & Schuster, 1971.
**McNeill,** William, *Plagues and Peoples*. London: Penguin Books, 1981.
**Meadows,** Jack, ed., *The History of Scientific Discovery*. Oxford: Phaidon, 1987.
**Meiggs,** Russell, *Trees and Timber in the Ancient Mediterranean World*. Oxford: Clarendon Press, 1982.
**Mennell,** Stephen, *All Manners of Food: Eating and Taste in England and France from the Middle Ages to the Present*. Oxford: Basil Blackwell, 1985.
**Nash,** Roderick, *Wilderness and the American Mind*. New Haven, Conn.: Yale University Press, 1967.
**Needham,** Joseph, *Science and Civilisation in China*. Vol. 5, Part 9 & Vol. 6, Part 2. Cambridge: Cambridge University Press, 1988, 1984.
**Pepper,** David, *The Roots of Modern Environmentalism*. London: Routledge, 1984.
**Perlin,** John, *A Forest Journey: The Role of Wood in the Development of Civilization*. New York and London: W.W. Norton, 1989.
**Pfeiffer,** John E., *The Emergence of Humankind*. New York: Harper & Row, 1985.
**Postgate,** Nicholas, *The First Empires*. Oxford: Elsevier/ Phaidon, 1977.
**Pounds,** Norman J.G.:
*An Historical Geography of Europe 450 BC to AD 1330*. Cambridge: Cambridge University Press, 1973.
*Hearth & Home: A History of Material Culture*. Bloomington and Indianapolis, Ind.: Indiana University Press, 1989.
**Pullan,** Brian, ed., *Crisis and Change in the Venetian Economy in the Sixteenth and Seventeenth Centuries*. London: Methuen, 1968.
**Reader,** John, *Man on Earth*. London: Penguin Books, 1990.
**Reischauer,** E.O., ed., *Ennin's Travels in China*. New York: Ronald Press, 1955.
**Rickman,** Geoffrey, *The Corn Supply of Ancient Rome*. Oxford: Clarendon Press, 1980.
**Roberts,** C. Neil, *The Holocene: An Environmental History*. Oxford: Basil Blackwell, 1989.
**Russell,** W.M.S., *Man, Nature and History*. London: Aldus Books, 1967.
**Salaman,** Redcliffe N., *The History and Social Influence of the Potato*. Cambridge: Cambridge University Press, 1949.
**Sauer,** Carl Ortwin, *Land and Life*. Ed. by John Leighly. Berkeley and Los Angeles, Calif.: University of California Press, 1963.
**Seymour,** John, and Herbert Girardet, *Far from Paradise: The Story of Human Impact on the Environment*. London: Green Print, 1990.
**Simmons,** I.G., *Changing the Face of the Earth: Culture, Environment, History*. Oxford: Basil Blackwell, 1989.
**Smith,** Bernard, *European Vision and the South Pacific*. New Haven, Conn., and London: Yale University Press, 1988.
**Spence,** Jonathan D, *The Death of Woman Wang*. London: Weidenfeld & Nicolson, 1978.
**Steven,** Margaret, *First Impressions: The British Discovery of Australia*. London: British Museum (Natural History), 1988.
**Stockwell,** Christine, *Nature's Pharmacy: A History of Plants and Healing*. London: Arrow Books, 1989.
**Tannahill,** Reay, *Food in History*. London: Eyre Methuen, 1973.
**Thacker,** Christopher, *The History of Gardens*. London: Croom Helm, 1979.
**Thesiger,** Wilfred, *Arabian Sands*. London: Penguin Books, 1964.
**Thomas,** Keith, *Man and the Natural World*. London: Allen Lane, 1983.
**Thoreau,** Henry D., *Walden*. Ed. by J. Lyndon Shanley. Princeton, N.J.: Princeton University Press, 1971.
**Tyler,** Ron, *Visions of America: Pioneer Artists in a New Land*. London: Thames and Hudson, 1983.
**Vainshtein,** Sevyan, *Nomads of Southern Siberia: The Pastoral Economies of Tuva*. Ed. by Caroline Humphrey. Transl. by Michael Colenso. Cambridge: Cambridge University Press, 1980.
**van Andel,** Tjeerd H., and Curtis Runnels, *Beyond the Acropolis: A Rural Greek Past*. Stanford, Calif.: Stanford University Press, 1987.
**van der Post,** Laurens, *The Lost World of the Kalahari*. London: Penguin Books, 1962.
**Walgate,** Robert, *Miracle or Menace? Biotechnology and the Third World*. London: The Panos Institute, 1990.
**Waters,** Frank, *The Book of the Hopi*. London: Penguin Books, 1977.
**Webster,** C.C., and W.J. Baulkwill, *Rubber*. Harlow, Essex: Longman, 1989.
**White,** K.D., *Greek and Roman Technology*. London: Thames and Hudson, 1984.
**White,** Lynn, *Medieval Technology and Social Change*. Oxford: Oxford University Press, 1962.
**Wolman,** M.G., and F.G.A. Fournier, eds., *Land Transformation in Agriculture*. New York: Wiley, 1987.
**Worster,** Donald:
*Dust Bowl*. New York: Oxford University Press, 1979.
*Nature's Economy*. Cambridge: Cambridge University Press, 1985.
*The Ends of the Earth: Perspectives on Modern Environmental History* (ed.). Cambridge: Cambridge University Press, 1988.
**Yap,** Yong, and Arthur Cotterell, *The Early Civilization of China*. London: Weidenfeld & Nicolson, 1975.

# INDEX

*Numerals in italics indicate an illustration of the subject mentioned.*

# PICTURE CREDITS

*The sources for the illustrations that appear in this book are listed below. Credits from left to right are separated by semicolons; from top to bottom they are separated by dashes.*

**Cover:** Photograph of oilseed rape fields in Fife, Scotland, Patricia Macdonald, Midlothian. **2, 3:** Robert Harding Picture Library, London. **8:** James Wellard/Sonia Halliday Photographs, Weston Turville, Buckinghamshire. **10, 11:** Jane Taylor/ Sonia Halliday Photographs, Weston Turville, Buckinghamshire; Trustees of the British Museum, London. **13:** Archivo Oronoz, Madrid. **14:** Trustees of the British Museum, London. **15:** Robert Harding Picture Library, London. **16:** Werner Forman Archive, London/Statens Historiska Museum, Stockholm; Horniman Museum, London. **18-19:** Art by Alan Hollingbery. **20, 21:** V. Southwell/Robert Harding Picture Library, London; Hirmer Verlag, Munich. **22:** Hirmer Verlag, Munich; Pelizaeus Museum, Hildesheim. **24:** Trustees of the British Museum, London. **25:** Werner Forman Archive, London/Schimmel Collection, New York. **26:** Map by Alan Hollingbery. **27:** Gianni Dagli Orti, Paris. **28, 29:** Roland and Sabrina Michaud/John Hillelson Agency, London; Erich Lessing, Vienna/ Trustees of the British Museum, London. **31:** Trevor Page/The Hutchison Library, London. **32-33:** David Burnett/Colorific, London. **34-35:** Marion Morrison/South Americaná Pictures, Woodbridge, Suffolk. **36-37:** © John Glover '79, Godalming, Surrey. **38:** Gianni Dagli Orti, Paris/Courtesy Musée du Louvre, Paris. **40, 41:** Elliott Erwitt/Magnum, London; Trustees of the British Museum, London. **42:** Scala, Florence/Museo Archeologico, Venice. **43:** Trustees of the British Museum, London. **44:** Erich Lessing, Vienna/National Museum of Damascus, Syria; Trustees of the British Museum, London. **45:** Trustees of the British Museum, London. **46:** Brooklyn Museum, New York. **49:** © Michael Holford, Loughton, Essex/Collection British Museum, London. **50:** Vatican Library, Rome – Trustees of the British Museum, London (3). **52:** Roger Wood, Deal, Kent. **54:** Trustees of the British Museum, London; Vatican Library, Rome. **55:** Trustees of the British Museum, London. **57:** Osterreichische Nationalbibliothek, Vienna. **58:** Collection of Tokyo National Museum. **60:** Paul Dix/Susan Griggs Agency, London. **61:** Courtesy of the Board and Trustees of the Victoria and Albert Museum, London. **62-63:** Réunion des Musées Nationaux, Paris. **65:** E. T. Archive, London. **66-67:** By permission of the British Library, London, OR 2362. **69:** By permission of the India Office Library (British Library), London, NHD 43 f.76. **70, 71:** © Wan-go H. C. Weng, Lyme, New Hampshire/Collection Palace Museum, Beijing; Rene Burri/Magnum, London. **73:** © Brian J. Coates/Bruce Coleman Limited, London. **74:** Museum of Fine Arts, Boston. **75:** Vatican Library, Rome. **77:** © Takeji Iwamiya/Pacific Press Service, Tokyo. **79:** Salgado/Magnum, London. **80:** Staatliche Museen zu Berlin; L. A. East, Liphook, Hampshire – Trustees of the British Museum, London. **81:** Staatsgalerie, Stuttgart. **82, 83:** Eileen Tweedy/ By courtesy of the Science Museum, London (2) – Ann Ronan Picture Library, Taunton, Somerset (2); The Board of Trustees of the National Museums and Galleries on Merseyside (Walker Art Gallery). **84, 85:** Drake Museum, courtesy of the American Petroleum Institute, Washington D.C. – British Nuclear Fuels plc, Warrington, Cheshire; Leif Berge, Hinna, Norway. **86:** Musée Des Arts Décoratifs, Paris. **88, 89:** Buckinghamshire Record Office, Aylesbury; By permission of the British Library, London, Roy 2B VII, ff112 and 155v. **91:** Bodleian Library, Oxford, Bodley 264, f81v. **92, 93:** Scala, Florence/ Biblioteca Marciana, Venice (4). **94:** Erich Lessing, Vienna/Graphische Sammlung Albertina, Vienna. **95:** Bibliothèque Nationale, Paris. **97:** By permission of the British Library, London, Harl.1585. **98:** Images Colour Library, Leeds; Bridgeman Art Library, London, by courtesy of the British Library, London. **99:** Bridgeman Art Library, London, by courtesy of the Kunsthistorisches Museum, Vienna; Hubert Josse, Paris. **100:** Fitzwilliam Museum, Cambridge. **103:** By permission of the British Library, London, Harl., 4425, 12v. **105:** Courtesy of the Board and Trustees of the Victoria and Albert Museum, London. **106, 107:** Trustees of the British Museum, London (2); Varga/Artephot, Paris; Antivarisk – Topografiska Arkivet, Stockholm – National Palace Museum, Beijing. **108-109:** Scala, Florence/ Ca'Rezzonico, Venice. **110:** Rijksdienst Beeldende Kunst on loan to the Frans Halsmuseum de Hallen, Haarlem. **112, 113:** Library of Congress, Washington D.C.; © Tom Bean/DRK Photo, Sedona, Arizona. **115:** Print Collection Miriam and Ira D. Wallach, Division of Art, Print and Photographs, New York Public Library, Astor, Lenox and Tilden Foundations. **116:** Trustees of the British Museum (Natural History), London. **117:** E. T. Archive, London; Eileen Tweedy/By permission of the Linnean Society of London – Trustees of the British Museum (Natural History), London. **118:** Trustees of the British Museum (Natural History), London; E. T. Archive, London/Trustees of the British Museum, London. **119:** Trustees of the British Museum (Natural History), London (3). **120, 121:** Robert Harding Picture Library, London/Collection National Galerie, E. Berlin; Archiv für Kunst und Geschichte, Berlin; Trustees of the British Museum (Natural History), London. **122:** National Gallery, London. **123:** Palace Museum, Beijing; Manchester City Art Galleries. **124, 125:** British Architectural Library, RIBA, London (2). **126:** Derby Museums and Art Gallery. **129, 130, 131:** Bridgeman Art Library, London/Collection Cheltenham Art Gallery and Museum; Tate Gallery, London. **133:** By permission of the British Library, London, 1038.b.24. **134:** By permission of the British Library, London, 45.i.3.4, p873; Mansell Collection, London – Photothèque du Musée de l'Homme, Paris; Mary Evans Picture Library, London. **135:** Scala, Florence/ Ca'Rezzonico, Venice; D. Graf/Staatliche Museen Preussischer Kulturbesitz Museum für Völkerkunde, Berlin – Trustees of the British Museum (Natural History), London – Copyright 1985 By the Metropolitan Museum of Art, New York. **136:** Sonia Halliday Photographs, Weston Turville, Buckinghamshire; Courtesy of the Board and Trustees of the Victoria and Albert Museum, London – Bodleian Library, Oxford, John Johnson Collection, De Bry folder I – Inventutis exercitia, pl. XXXVI. **137:** Peter Newark's American Pictures, Bath – Royal Botanic Gardens, Kew; By permission of the India Office Library (British Library), London, NHD 19, f8. **138:** Courtesy of the Board and Trustees of the Victoria and Albert Museum, London; By permission of the India Office Library (British Library), London – By permission of the British Library, London, 447.g.4, p252; Mary Evans Picture Library, London. **139:** Wellcome Institute Library, London – Royal Botanic Gardens, Kew, London – National Library of Medicine, Bethesda, Maryland; By permission of the British Library, London, 451.i.3.4, p419. **140:** By Ralph Perry © National Geographic Society, Washington D.C. **142:** By permission of the British Library, London, 713.i.1.11 F/P. **143:** Museum Folkwang, Essen. **144:** Trustees of the British Museum (Natural History), London. **145:** Courtesy of the Kendall Whaling Museum, Sharon, Massachusetts. **146:** Foreign and Commonwealth Office, London. **147:** © 1990 Monas Hieroglyphica, Milan. **148, 149:** By Steve McCurry © National Geographic Society, Washington D.C.; Steve McCurry/Magnum, London. **150, 151:** Archivio di Stato, Venice (3); National Museum of American Art, lent by the US Department of the Interior/Art Resource, New York; G. Ziesler/Bruce Coleman Limited, London. **154, 155:** By David Turnley © National Geographic Society, Washington D.C.; Map by Town Group Consultancy. **156:** Sebastiao Salgado/ Magnum, London. **158:** Steve McCurry/ Magnum, London – Charlotte Raymond/ Science Photo Library, London. **159:** Hank Morgan/Science Photo Library, London. **162-163:** Peter Menzel/Science Photo Library, London. **164:** Greenpeace, London – Friends of the Earth, London – World Wild Fund for Nature, Godalming, Surrey – Greenpeace, London **165:** NASA, Goddard Space Flight Centre Greenbelt, Maryland.

# ACKNOWLEDGEMENTS

The following materials have been reprinted with the kind permission of the publishers: Page 9: "Here every man. . ." quoted from *Arabian Sands* by Wilfred Thesiger, London: Penguin Books, 1964. Page 18: "For about two minutes. . ." quoted from *The Lost World of the Kalahari* by Laurens van der Post, London: Penguin Books, 1962. Page 62: "The people live. . ." quoted from *The Songs of the South: An Anthology of Ancient Chinese Poems,* transl. by David Hawkes, London: Penguin Books, 1985. Page 68: "Ten tubes upon each wheel. . ." quoted in *The Pattern of the Chinese Past* by Mark Elvin, London: Methuen, 1973. Page 75 and *passim*: "destitute and ravaged. . ." quoted in *The Death of Woman Wang* by Jonathan D. Spence, London: Weidenfeld & Nicolson/Peters Fraser & Dunlop Group, 1978. Page 87 and *passim*: "After men had eaten. . ." quoted from *The Five Books of Histories* by Radulfus Glaber, ed. and transl. by John France, Oxford: Clarendon Press, 1989.

The editors also wish to thank the following individuals and institutions for their valuable assistance in the preparation of this volume:

**England:** London – Michael Donaghy; Anne-Marie Ehrlich; Rupert Hastings, The Royal Botanic Gardens, Kew; Beth McKillop, Oriental Collections, British Library; Christine Noble; Andrew Norris, RIBA Drawings Collection; Chris Rawlings, Photographic Services, British Library. Oxford – Suzanne Williams.

**U.S.A.:** Sharon, Mass. – Stuart M. Frank, The Kendall Whaling Museum.

The index for this volume was prepared by Susanne Atkin.

Typeset by A.J. Latham Limited, Dunstable, Bedfordshire, England.
Colour separations by Fotolitomec, S.N.C., Milan, Italy.
Printed and bound by Brepols S.A. — Turnhout, Belgium.